Geographical Sociology

The GeoJournal Library

Volume 105

For further volumes:
http://www.springer.com/series/6007

Jeremy R. Porter • Frank M. Howell

Geographical Sociology

Theoretical Foundations and Methodological Applications in the Sociology of Location

Springer

Jeremy R. Porter
Brooklyn College & Graduate Center
City University of New York
218 Whitehead Hall
2900 Bedford Avenue
New York, NY, USA

Frank M. Howell
Emory University
Atlanta, GA, USA

Mississippi State University
Starkville, MS, USA

ISSN 0924-5499
ISBN 978-94-007-3848-5 e-ISBN 978-94-007-3849-2
DOI 10.1007/978-94-007-3849-2
Springer Dordrecht Heidelberg London New York

Library of Congress Control Number: 2012933991

Printed on acid-free paper

Springer is part of Springer Science+Business Media (www.springer.com)

To Linda, Jason, and David.

JRP

To Jon D. Holmes, DMD, MD, FACS,
the surgeon who saved my life.

FMH

Contents

Chapter 1
Geo-Sociology

1.1 Geo-Sociology Defined

Recently, increased attention has been given to the relationship between geographic context and the ecological settings in which social process and behaviors of both groups and individuals occur. This focus has been inherently geographic in its focus on concepts associated with such examples as individuals being situated within a neighborhood or populations in proximity to a natural disaster. We argue in this book that while the methodological tools for such analyses are relatively new in their development, spatial thinking has long been at the core of much of the traditional sociological theory that marks the foundations of the discipline. Together, these theoretical foundations, coupled with the more recent methodological ability to put "people into place", makeup an emerging movement towards a geographically focused sociology. In its most generic form, and on the tail of a recent publication (Porter 2011), *geo-sociology* should be viewed as a synergy between ecologically-centered macro theory and the application of spatially-centered research methods in the examination of sociological questions. This informal definition is useful in the sense that it identifies the role of geo-sociology, in both practice and principle s, and in regards to the complimentary role that theory and research play within its developmental framework.

We make no claims to be the first to introduce the components of such an approach, only to contribute to this development by introducing a unifying medium through which the historical foundations, contemporary understandings, and future developments of geo-sociology can be better understood. Historically, Edward Hayes (1908) first wrote of the need for a geographical sociology. Hayes' 'geographical sociology' was one which aimed to understand the role of the physical environment as an explanation for variations in human behavior and population level processes. Here we move beyond Hayes' initial introduction of the geographic sociology by linking over 100 years of theoretical and methodological developments in the formation of a modern geo-sociology. Ultimately, this attempted unification of theory and methods sets geo-sociology apart from the many existing, but

J.R. Porter and F.M. Howell, *Geographical Sociology: Theoretical Foundations and Methodological Applications in the Sociology of Location*, GeoJournal Library 105, DOI 10.1007/978-94-007-3849-2_1, © Springer Science+Business Media B.V. 2012

fragmented, approaches to the examination of human behaviors and population processes in their given environmental context from which we draw upon heavily.

As evidence of the current levels of fragmentation in this focus of sociological research, a major portion of the text that follows is inspired by the historical disjuncture in the early development of social ecological theory and the technological advancements in data-availability and computing that have driven the resurgence of macro-level sociology. Drawing on this existing knowledge, and building on our own understanding of the current trajectories of such an approach, our argument is that the development of the macro-level theories that provided the earliest foundation for the disciple of sociology itself did not happen in historical unison with the more recent capabilities to empirically test such tenants and thus continue to remain distant. By reflecting on the historical development of geo-sociology, we hope to push forth future discourse that would continue to apply the approach presented in this text to the practice of ecologically-centered sociological research.

It is important to note that we situate geo-social approaches to specific research endeavors within the same framework as other, more recognizable, approaches to "doing sociology" such as the psycho-social approaches that have stood the test of time and the bio-social approaches that are proving to be increasingly important as the lines between fields such as social science and medicine continue to blur. In summary, this book is dedicated to the discussion of geo-sociology as a concept within the larger discipline of sociology. In doing so, we document the historical development and future trajectory of geo-sociology as a viable approach to understanding human behaviors and social processes but with a current disjuncture in theory and method both within the larger field of sociology and in relation to recent ecologically centered social research.

1.2 Sociology: A Discipline of Syntheses and Contexts

In order to understand the emergence of geo-sociology, we must first understand the propensity for evolutionary development within the field sociology. Such development is historically related to the alliance of sociological thought to cross-disciplinary research approaches. Early in the last century, such a Kuhnian approach to evolutionary science was articulated with a call for its continuance as an avenue to reach the fullest potential of sociology as a discipline. In *Recent Developments in the Social Sciences* (1927), Charles Ellwood of the University of Missouri outlined some early developments in the field of sociology. Two major points Ellwood (1927) introduced were (1) the drive to move beyond 'organic' or 'synthetic' understandings of social phenomena and (2) the accomplishment of this through the synthesis of several cross-disciplinary analytic methods. The first point was a call to better understand the mechanisms driving many of the social processes that were identified by sociologists in the day. The second point is much more important to the subject of this book, the development of geo-sociology as a synthesis of some social theory and geographical methods.

At the time of Ellwood's survey, the synthesis of social theory and non-native (to Sociology) methods were to be charged with ultimately producing a method of its own aimed at developing an over-arching understanding of society, through the combination of borrowed methods from sister disciplines. At that time, the most influential disciplines within the then emerging field of American sociology, were psychology and biology, which over time have both developed into strong concentrations within the field of sociology. The ability to place individuals within specific psychological and biological dimensions became important in ultimately allowing researchers to better understand individual action in terms of a given context (Ellwood 1927).

This idea of context, and its conceptualization, has also evolved along with the discipline of sociology. In fact, there is an inherent drive of sociologists to understand the context in which human action occurs. In many cases, landmark research has continually pushed along our understanding of human behavior based on locating individuals within a given contextual frame. It is important to note that this frame does not necessarily have to be physical. As many are familiar with the concepts social standing, race, time period, and so forth, all of these represent a contextual frame in which human action occurs. One of the most notable in this line of research is the synthesizing work of C. W. Mills (1959) in his book, *The Sociological Imagination*. In this essay, Mills emphasized the importance of contextual biography to the understanding of human agency within given structural situations saying that, "*No social study that does not come back to the problems of biography, of history, and of their intersection within a society has completed its intellectual journey*" (p. 5). Since that time, many have extended and worked within the context of the *Sociological Imagination* as a framework for understanding unique individual circumstance (for example see Shanahan and Macmillan 2008; Fuller 2006).

It is difficult to deny that the quote from Mills is powerful. It speaks to the power of context and the inability to separate the understanding of human behavior from such contextual situations. More importantly, this continued the standard in sociology that was associated with the interdisciplinary pattern of drawing on established disciplines. In this case, Mills brought to light the continued development of sociology by coupling the theory and methods of history and sociology for the development of a subfield such as historical sociology. Within a historical framework, the increasing intersection of society's development further allowed for the continued emphasis of contextual location as a determinant of individual behavior. However, as Abbott (1994) notes, historical sociology has not seen the same representation as many other sub-fields in sociology and he has referred to it thusly as "The Lost Synthesis". However, its simple development, in combination with other contextually driven approaches to understanding society, speaks to the ever-increasing role of context in sociology.

A similar pattern has occurred regarding the inclusion of "space" in sociological research. However, as this book documents, the re-emergence of "space" in sociological research has largely occurred in unison with advancements in geo-social methods of quantitative data analysis.

The synthesizing nature of sociology debatably represents a following of the patterns identified by Thomas Khun (1962) in *The Structure of Scientific Revolutions*.

Khun (1962) states that the development of science is historically related to the synthesis of competing and complimentary theoretical epistemologies in the face of scientific anomalies. It can be argued that the marrying of sociology to biology, to psychology, and to history (… etc.) is a consequence of such anomalies encountered by early sociologists. The application of these complementary theories and methods has allowed for the development of a dynamic and more meaningful discipline of sociology that we see today. On that point, the inter-dependence is not a sign of weakness but of progress towards a "better sociology" through the continued development of paradigm-directed analytic methods (Kuhn 1962). It should be expected that, as long as the discipline continues to draw upon other related disciplines in the quest to better understand social phenomena and processes, it is continually progressing towards a "better sociology" (Levine 2004).

When relating the progress of a discipline to a point of origin, some researchers have professed a belief that the roots of sociology, and the social sciences in general, arose much earlier than has been formally acknowledged (see McDonald 1993), while still others acknowledge the industrial age as a seminal point (see Olson 1993). What cannot be argued is that from either viewpoint, the historical development of a contemporary sociology is a manifestation of a synthesizing, often reactionary, and interdisciplinary approach to the examination of human and group action embedded within the environment in which they habitat. The environment in which they exist being *the* key concept in this text and relating to the more general term of context, which, not by coincidence, finds its beginning in the various syntheses outlined above.

1.3 Sociology and Geography: A Synthesis of People and Place

One synthesis, which has until recently received relatively little formal recognition, is geography and sociology. Yet, thc discipline of sociology, and its general definition as the study of societies, institutions and human interaction, has an undeniable linkage to space. Societies exist in geographic places on our earth and the social institutions, which are developed and maintained within those societies, are governed by both informal (mores, norms, values, etc.) and formal (laws, social/economic policy, etc.) directives that often do not cross imaginary boundaries on the physical earth. Furthermore, we can all recall some of the earliest works to be considered sociology were developed in response to rapid and massive social change occurring in urban places as a result of industrialization (for instance Park and Burgess' 1925 work *The City*). Thus, the relevance of place within sociology is inherently clear. Perhaps the most widely accepted difference between the disciplines of sociology and geography was best presented by Robert Park (1926) when he made the point that geography is concerned with the atypical and sociology with the typical. Thus, while geographers, even human geographers whose research is speaks very well to sociological research, are interested in deviations from patterns, sociologists are generally interested in uncovering patterns. Both however, are interested in research across populations that are situated proximally in relation to one another and in relation to the available land, resources, and physiographic features that mark the geographic landscape. Sociology has always held geography as a central focus in its research, whether it was intentional or not. In relation to other closely aligned social sciences (i.e. political science and economics), which also have an interest in aggregate groups, sociology has historically been a the forefront of incorporating space into its research.

As evidence, a quick Internet search engine request, in the bibliographic database JSTOR™, for the key word "context" shows that the synthesis has been noticeably embedded within the discipline, especially in the very recent past and in relation to other social science disciplines. From Fig. 1.1, there is the total number of occurrences of terms in relation to concepts of space presented in the top panel while the proportion of all publications containing such words are presented in the bottom panel. In regards to the total number of publications in which these key words were found in the title of the publication, they were slightly more prevalent prior to the late 1950s. Post 1950s, their prevalence grew at a faster rate than that of Economics or Political Science but by the 1980s they seemed to be appearing at equal rates. Finally, since about the mid-1990s the usage of these terms in the titles of publications in sociology has once again grown in comparison to political science, which continues to embrace space at a slower rate, and economics, which has actually seen a decline in the usage of these terms in recent years.

While the top panel provides some evidence of the higher importance of geography in sociology, when compared to economics and political science, the bottom panel presents these findings in standardized form as a proportion of all articles published by year. By focusing on the trends, one can see that historically sociology had a higher proportion of articles related to terms of ecology, space, place, location and geography. During the 1980s this proportion dropped. Since the early 1990s, however, the trend suggest that sociology is "recapturing space" as an important determinant within the research being published. It should be noted that these publications only make up a small proportion of the total publications identified, however there are many other search terms that relate to ecological research that were not included in this analysis and would also be expected to provide further support of sociology as leading the way in research concerned with the spatial proximity of the populations being studied to their specific geographic context.

Not surprising, the role of ecology, geography, location and space in sociological analyses are most often reserved for those whose substantive interest more directly ties them to space, such as work by urban or rural sociologists, demographers, criminologists, and epidemiologists. In a larger sense, all are distinct, yet fundamental, parts of the sociology enterprise yet each has a direct link to the spatial relationships of macro-level units of analysis, displacement, diffusion, epidemics, and other spatial (and spatio-temporal) relationships. What is somewhat surprising, given the lack of formal recognition mentioned above, is its corresponding growth of spatially-centered theoretical and analytical approaches in mainstream general sociology journals. While it is somewhat expected given its increased visibility and

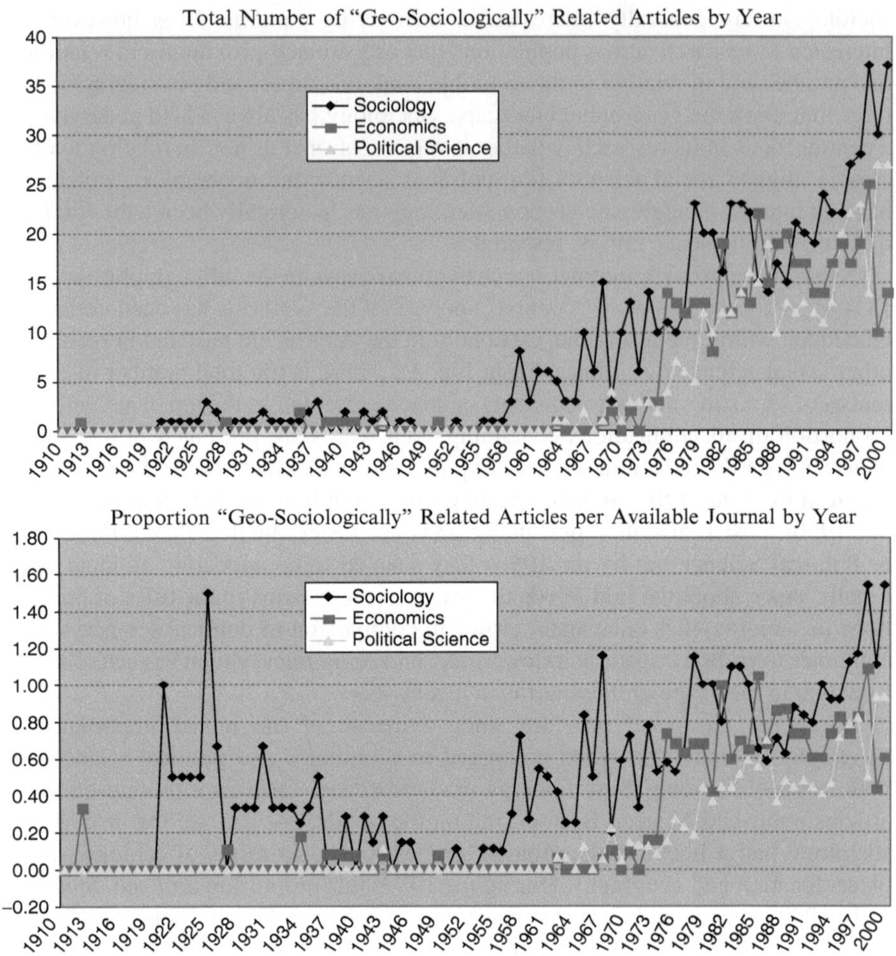

Fig. 1.1 Total number and proportion of articles identified as spatially-centered in analytic methodology or substantive interest (*Data Source*: J-STOR Social Science Journals Search Engine and Author's Calculations. Search criteria included words such as ecology, location, context and place)

usage in sub-disciplines of sociology comprehensively, little is known within the field about its historical development, present applications, and future trajectory of geo-sociology. Furthermore, what has been developed is currently compartmentalized as spatial analysis, ecological modeling, and similar approaches.

Geo-sociology refers to the unique situation of social behaviors within the ecological confines of a given geographic landscape. Some sociologists, like Wilkinson (1991), call this locality. It is related to geography in its ability to descriptively locate population characteristics in space. However, the methodological role of geo-social analyses moves beyond this important stage to a more dynamic position involving weighting schemes, distance analysis, and explanatory power. That is not to say that geo-sociology is solely quantitative. As with all analytic approaches in the field of sociology, it finds its beginnings in a well grounded series of cross-disciplinary theoretical frameworks pulling from the closely related disciplines of geography, economics, sociology, and others.

Again, the purpose of this book is to introduce the concept of *geo-sociology* as sub-discipline of sociology, drawing heavily on geographic methods and spatial analysis but directly linked to theories of place and the role that space plays in directing human behaviors and population processes (i.e., social demography). We readily concede that much of our work here is the bringing together of much material on the subject which already exists but has yet to be organized in any meaningful fashion. We do this through a short historical review of the progression of our understanding of society and space. This is undertaken through a chronological examination of cross-disciplinary work using spatial concepts followed by an overview of recent developments in the analytic methods that have, both directly and indirectly, contributed to the rise of the current attention given to these analyses in sociology. In many ways, the recent developments in spatial analytic methods have allowed sociologists to re-visit the historic core tenants of the discipline and how they may aid in the understanding of the direct causal effects of varying ecological conditions on human behavior. In fact, while aggregate patterns exist in society, not explicitly controlling for one's spatial propinquity to a given ecological situation decreases our understanding of the confounding effects associated with geographic context.

1.4 Book Organization

We introduce many of the historically relevant theoretical perspectives related to the development of a geo-sociology in the first few chapters. As readers of this book will see, the roles that theories of space and place in the development of sociology have been more than simply important. For instance, drawing on early economic theories of location and space, some of the first American sociologists to pay attention to the importance of this relationship were researchers at historically influential institutions, such as the University of Chicago and the rural sociologists at University of Wisconsin at Madison. While these conceptual frameworks were usually hindered at the time by a dearth of available quantitative methods for analyzing such relationships at larger scales, this early work was essential in laying the ground work for the importance of spatial-location theory in the understanding of social behavior.

We also discuss the recent technological advancements in the field that have allowed for the exploration of tenants of space, place and geography within the larger field of sociology that were first posited by this early research. Finally, we end with a prospectus concerning the future trajectory of geographical analysis in sociology. While the discussion that lies ahead in this text is certainly not exhaustive in form, we hope that it introduces some of the more influential concepts, theories and methods in the historical development of a geo-sociology.

We have organized the book into two sections in hopes of bringing together the wealth of existing information regarding geo-sociology and its roots, development, and place in the larger discipline of sociology.

The first section of the book is an overview of the historical development of geo-sociology. This section includes Chaps. 2, 3, 4, and 5 and focuses on the important milestones in the historical development of theoretical and methodological approaches to understanding the role of place and its impact on human behavior. Furthermore, the chapters in this first part of the book are directly related to some of the seminal theoretical explanations which tie people to place. This first section includes an initial chapter presenting a discussion of the early theories of place linked to economic efficiency and the development of the city. Following that is a more sociologically relevant discussion outlining the role of space made famous in the discipline by urban/community sociologists in the Chicago School of Sociology and the unrealized impact of the Rural Sociology program at the University of Wisconsin which predates it. Finally, two chapters focusing on a more contemporary discussion of human ecology and environmental sociology conclude the section.

The second section of the book presents the current state of geo-sociology in the study of societies and the people and institutions which make them up. In the chapters making up the second part of the book we bring together information regarding recent advances in technology that has led to the rapid growth of research in sociology which focuses on the role geography and place. These chapters explicitly focus on the ability to test tenants and propositions put forth in the preceding section of the book that focused on the historical development of geo-sociological theory. Ultimately, this section highlights the temporal disjuncture in the development of theoretical propositions and the more recent ability to actually test and build upon those propositions with the advent of technology designed for sophisticated spatial analyses. In doing so, this section begins with a chapter on linking social data to GIS and other software that allows for the placing sociological research in geographic context. The following chapter highlights methodological advancements, as well as their development, to the development of geo-social theory presented in the first section of the book. This should be considered one of the more important contributions of the book as it provides links between why we think geographic ecology impacts human behavior (geo-social theory) and how we put those theories into practice (geo-social methods). To date much of the linkage between geo-social theory and geo-social methods is relational, relying heavily on theory outside of the discipline of sociology (usually geography) or absent in the sense that much of the theory driving the current boom in research interested in the "neighborhood-effect" is atheoretical with a with little development of theories aimed at addressing *why* geographic context effects individual behaviors and group processes. Finally, the last chapter in the second part of the book provides a set of empirical examples of geo-social analyses using real-world social data that concerns issues of general interest to sociologists.

Our goal in the development of this book was to bring together the wealth of scattered materials that contribute to geo-sociology and present them in an organized and accessible fashion. We hope that the chapters that follow lead to further discussions concerning the increasingly relevant place that geo-sociology plays in the larger discipline of sociology. For as we show, the propositions and tenants of s geo-sociological theory have long since been present and the analytical abilities to "put theory into practice" are now available and accessible to a large enough group of researchers to make them meaningful to the discipline at-large.

Part I
The Sociology of Location: Theoretical Foundations

Chapter 2
Important Contributions to Geo-Sociology

A number of location-specific spatial theories of human and societal organization have developed over the past century. As mentioned in the previous chapter, the development of spatially-centered analytic research methods, aimed at understanding the organization and behaviors or individuals and populations in space, has not occurred in unison with these disparate bodies of extant theory. To put this relationship into perspective, theories of location and population organization have long been formally documented, complete with supporting facts and well-developed theoretical propositions. However, the ability to actually test many of these spatially-focused propositions has only recently become widely available. Recent catalysts include the ubiquity of the personal computer and the computational power that has accompanied this technological advancement. Following this, the emergence of spatial software to harness such computing power followed slowly along with the cheaper personal desktop computers of the past decade or so. As such, the theoretical development of theories of location is well established and traces its roots across many different disciplines. We have not had a cogent assemblage of them in one place to facilitate integrated thinking about them in a coherent way.

In some of the earliest theories the understanding of the role of place and its central nodes was the primary focus. For instance, in Von Thunen's *Isolated State* (initially published in 1826, citations refer to 1966 edited English version), the primarily purpose was to understand the economic geography underlying the physical situation and development of the community. In this sense, the community itself was self-sufficient and developed solely on the relationships between economic costs associated with both commodities and transportation of goods to the market. Ultimately, this model was deemed too simplistic but the resulting ring-based model that has come to represent the foundation for many important developments in location theory that followed.

Subsequent theories focused on the *interdependencies* that developed between communities in space and the natural hierarchy that occurred as a result. For instance, Christaller's theory of Central Places (translated in 1966) was one such perspective that was explicitly interested in the development of "places" in a planar

J.R. Porter and F.M. Howell, *Geographical Sociology: Theoretical Foundations and Methodological Applications in the Sociology of Location*, GeoJournal Library 105, DOI 10.1007/978-94-007-3849-2_2, © Springer Science+Business Media B.V. 2012

space and the differential development in magnitude given the concentration of supply and its relative relationship to demand. Closely related is Weber's (1909, translated in 1929) theory of the location of industries, aimed at identifying the optimal economic location of industry during the time of rapid industrialization.

Other important contributions include the work of August Losch, Charles Booth, and John Snow. Losch's *The Economics of Location* (originally published in 1939, but citations refer to the 1953 translation) and *Population Cycles as a Cause of Business Cycles* (1937) which seminally contribute to our understanding of *regions* as economic units and the relationship between population and production. Famously, John Snow's *On the Mode of Communication of Cholera* (1949) was one of the earliest theories to be developed through a crude method of mapping an 1854 cholera outbreak in the city of London. Similarly, Charles Booth's voluminous *Labor in the Life of People* (first volume published in 1889) was an attempt to understand the spatial distribution of poverty in the city of London.

Other important contributions have been made over the years since then but are too plentiful to be included in the scope of this book chapter. At the time of publication, the most complete set of "classics" in the development of our current state of location theory known to the authors is the website of the Center for Spatially Integrated Social Science (or CSISS; see www.csiss.org). The CSISS has not only proven to be an important source of information concerning the documentation of contributions to the use of location-specific theories in the social sciences but also important in its own unique contributions which will be introduced in a later chapter. The remainder to the current chapter is dedicated to a more in-depth understanding of some of the more important (arguably of course) pieces in the existing literature that have contributed to the historical development of a geo-sociology as they have been briefly introduced above.

2.1 Early Roots

The origins of locational-centered spatial theory in the social sciences lie firmly rooted in the fields of geography and economics. In fact, if one were to survey the landscape of introductory theory courses in either geography or economics, they would very likely be introduced to much of the early research in this area. If these theoretical frameworks were not explicit, then most definitely one would find vestiges of these early contributions underlying much of what we consider to be the current state of the art in theories of location. The timing of such developments should also not be lost on readers of this text. From the mid-nineteenth century through the early turn of the twentieth, cities began to be the dominant form of organized society. At this time, the industrial revolution had firmly taken hold and produced a market-driven society in which the ability to manufacture/produce commodities, transport to market and sell at a rate equal to demand was the emergent model of business.

Given this model, some of the earliest theories of location were derived specifically in order to address the maximizing of profit. Thus, the developments of economic geography perspectives were concerned with distance to market and the organization of societies within the ever-growing cities. In these models, the ideas represent logical use of space or location as a way to maximize earning potential, whether through land use in agriculture or the development of urban areas in the analysis of a hierarchy of places (cities and towns). Along the way, other important developments have occurred and contributed to the continued development of spatial theory, such as the development of a similar model to the concentric model of Thunen Model by Park and Burgess in their landmark publication *The City* (1925). Within this work, the concentric zonal model was introduced by researchers from the Chicago School as a further development of research from the seminal work of Galpin (1915) in his *Anatomy of a Agricultural Community.* As a whole, these developments have furthered the range of spatial theory and have helped developed it into one of population and people, therefore transforming this traditionally economic theory into one of importance to social scientists, demographers, and economic theorists alike. Arguably, one of the most influential contributions in the development of a geo-sociology is the development of *central place theories*. These theories were among the first to tie individual behaviors to spatial locations.

2.1.1 Von Thunen's Zonal Model of Proximity and Economic Maximization

One of the earliest theories of central places and spatial location in economic geography is that of J. H. Von Thunen (1783–1850), who was a German farmer and amateur economist. In Von Thunen's *Isolated State*, a zonal model of marketplace, landuse, and transportation was introduced (1826) and all following references to this model refer to the 1966 translation (edited by Peter Hall). The model introduced by Von Thunen is concerned with the spatial arrangement of various land covers to the central city and their systematic arrangement as a result of maximizing the potential profits of farmers in the community. It is important to note that this model was developed during the earliest onset of the industrial revolution and the associated development of factories and, as a result, this model is based on a number of limiting assumptions. Furthermore, Von Thunen himself was a farmer first and an economist second, thus his interest in the ability of farmers to grow and maximize profits. A simple presentation of the model is presented as a concentric model in Fig. 2.1.

As a result of his background and primary occupation in farming, the Von Thunen model was applied to farms as a way of planning the planting of crops based on their harvesting and transportation costs. The model is economically grounded in hopes of balancing land costs with transportation costs and is simple in structure as there are four rings of agricultural activity surrounding the central city. Of importance to note, the model has been developed around the idea that the community (or city) is self-sufficient and there interdependencies between the community and neighboring communities. Thus, the community operated in an "isolated state". Furthermore, the model's concentric shape is predicated on the assumption that no transportation routes exist and that any one point in the spatial plane is the same as another other

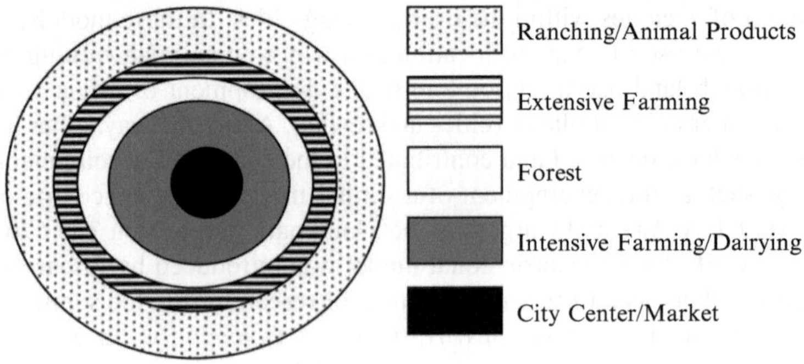

Fig. 2.1 J. H. Von Thunen Landuse/transportation cost organization model

point given their equality in distance from the center of the model (market place). Thus, distance is the only variable that comes into play in the develop of Von Thunen's model and all other assumptions are predicated on the location of specific goods and tasks as a result of balancing costs due to the maximizing of profits from distance constraints. Furthermore, the land is assumed to flat and devoid of waterways, the soil consistent throughout the plane, and farmers are responsible for the delivery of their own crop to the marketplace in order to make sales.

What is most remarkable about this model is that many of the model's assumptions tended to fit the development of early cities (see specifically the concentric zonal models of *The City* in the next chapter). Ultimately, all of that changed with the advent of transportation and communication technologies but the foundation that was laid by the model continues to exist in many modern theories of location. The model itself is designed so that the innermost ring (directly outside of the marketplace) is where intensive farming takes place in order to reduce transport costs to the city. In this ring, produce such as lettuce, cauliflower and strawberries would be farmed as that that time they would be unable to last for long journeys to the market. Furthermore, milks and cheeses were likely to be farmed in these areas due to their high demands and high likelihood of also perishing on long journeys. Von Thunen makes it a point to explain that no land in this ring would ever go bare for a couple of reasons. First, its proximity to the marketplace is optimal, thus driving up land rents, secondly, due also to proximity to the communities center, manure would be widely available and drive up the quality of the soil in this region. Once beyond the first ring the price of fetching manure and taking products to the market increase too high to make them profitable.

The next ring consists of timber and firewood for fuel and building. Its placement was also determined by the fact that timber was heavy and hard to transport. It is important to remember the time at which this model was developed. During the late 1700s and early 1800s timber was the primary source of fuel and shelter for all residents of the "isolated" community in which Von Thunen's model would have applied. Thus, it makes sense that the proximity of such a valuable resource would only be second to perishable items that could not make the trip. However, in relation to less perishable items such as grains, the weight of the timber makes it economically less rationale to grow far from the town's center. In this sense, the Euclidian transportation distance becomes a factor associated with the effort required to move the commodity to the marketplace and maximizing the potential profits associated with that transportation and sale.

Directly outside of the ring of forest and timber is the third ring which consists of extensive farming crops. Extensive field crops, which include grains, occupy the third ring because they keep longer than dairy and are much lighter than fuel. Once again, the transportation costs are directly tied to both distance and effort in transportation. Given, the relative ease of transportation, when compared to fuel (timber), grains would be grown outside of the region where timber was grown in order to maximize profits. Von Thunen points out that timber and grain have no compatible use value and so when understanding the proximal location of their farming to the town's center one must take into account the conditions that have been mentioned here. First, the timber is much heavier and therefore would take much more effort to move to the market and secondly, the non-perishable state of grains would allow them to make the much longer trip from this ring in Von Thunen's model. Furthermore, as the land rent moves toward zero, the cost of producing the grain is almost all tied up in transportation. Thus it makes the most logical sense to move something lighter and easier to transport from these out regions.

Finally, the last ring in the model is reserved for ranching/animal products, which can be raised far from the city because they are self-transporting. Von Thunen's model estimated that this ring would begin at about 30 miles from the town's center. With the ignorable cost of land rent and the little cost or trouble associated with moving ones livestock from place to place, the out ring of the model is the most rational for stock farming. Furthermore, from a purely comparative point of view, when looking at the price of land and the effort associated with transporting commodities one self, no other areas in the model are able to compete with this outer ring as the most efficient location for stock farming. Also, as land rent approaches zero much more arable land is available for pasture, which provides the primary source of nutrition for most of the livestock.

From this model and the accompanying assumptions, it is clear that these ideas are very primitive and, taken as is, they are of almost no use to modern spatial theory. In fact, the idea that there is no variation in land elevation, water supply, soil fertility, transportation veins, etc. is archaic in the least. In fact, without creativity, it would be easy to dismiss this model as archaic and inapplicable. Of great importance, however, is the fact that these ideas provided the initial foundation on which other ideas were built. Much of the subsequent theories of central places and location of industry theories were derived from a starting point associated with Von Thunen's model of land use. In these later models, many of the assumptions are filled in along the way however, for many years since its introduction the distinguishing concentric zones associated with Von Thunen's model remained in much of the theoretical contributions concerning location and the organization of cities in space.

2.1.2 Christaller's Central Place Theory

In 1933, Christaller's Central Place Theory (CPT) was published using many of the same ideas introduced by Von Thunen over a century earlier (Haggert 1967). CPT, like the Von Thunen Model, was grounded in economics. It is important to note that while Christaller is most often linked to the development of CPT, the idea of "central places" actually appeared in publication form 2 years prior to Christaller's (1933) publication of *Die Zentralen Orte in Suddeutshland* (translated: *Central Places in Southern Germany*) (Marshall 1969). In fact, Mark Jefferson's (1931) "Distribution of the World's City Folks", was published in *Geographical Review*, and is first credited with summarizing the hierarchical relationships among cities which makes up the core of CPT. Ultimately, Edward Ulman introduced Christaller's work in English a decade after its conception (Marshall 1969). The following summary of CPT is in reference to that work which was published by Ulman in the *American Journal of Sociology* as "A Theory of Location for Cities" in 1941.

As mentioned above, the foundation that was Von Thunen's concentric zonal model provides the basis for CPT. However, unlike Von Thunen's model, CPT realized that the city is neither isolated nor is it self-sufficient, and, based on that idea, Christaller developed a hierarchy of cities or towns, in which cities are larger in both magnitude (size) and diversity associated with the production of essential and non-essential goods. Furthermore, the spacing of cities was essential to the development of the hierarchy associated with places in Christaller's CTP.

This hierarchy was developed based on two basic concepts; (1) threshold-the minimum market needed to bring a goods seller into existence and keep it in business and (2) range-the maximum distance people will travel to purchase goods. The range, then, was further spatially divided based on lower or higher order goods; lower order goods were those consumers need less-frequently and, therefore, are less likely to travel long distances for them and higher goods which consumers are likely to travel for given the frequency of use. Without getting into to much detail, the price of goods is directly related to both the availability of willing buyers (threshold) and the distance in which people are willing to go to purchase it (range). If a product is too expensive, then in order to increase the threshold and seller much lower the price to get people to give buyers a reason to travel further (increasing the range) to buy the product, thus increasing the threshold. Again, we see the goal of the economic maximization of profits in the most basic foundations of CPT. The most basic presentation of CPT has been graphically presented in the form of the diagram in Fig. 2.2.

From the basic diagram in the top panel of Fig. 2.2, one can see the most basic tenant of Christaller's CPT. The theory is directly concerned with a central trade center and the activity that disperses from that center. The threshold and the range are directly related to one another, primarily depending on the goods being sold. In this most basic diagram, the central trade center is demarcated by a black circle and the marketplace it serves is an empty larger surrounding circle. In this model, each supplier has their own market place and no competition exists amongst them.

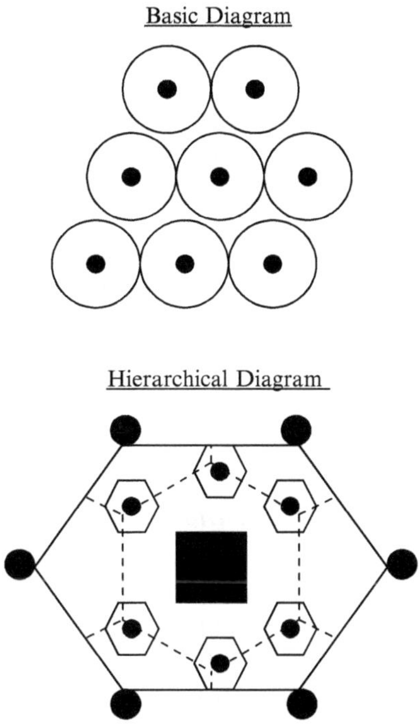

Fig. 2.2 Christaller's central place theory: basic and hierarchical diagram

Christaller's theory adds to this basic diagram by asserting that given the threshold of any city certain commodities exist in a hierarchical fashion. This hierarchical diagram is presented in the bottom panel of Fig. 2.2. In larger central trade centers (large cities, represented by the large square in the diagram), which are more likely to have higher order goods, the range would be much further than the threshold as people are willing to travel further for higher order goods. In the smaller trade centers (smaller towns, represented by the small circles in the diagram), the range is not very big as they are more likely to have only lower order goods and, as mentioned earlier, people are not willing to travel very far for lower ordered goods. Ultimately, as you move away from the large city center you begin to develop medium size towns and cities with smaller ranges and thresholds that the large urban cities, but larger thresholds and ranges than the small towns and villages (represented by the larger circles on the outside of the diagram). This, then, sets up a hierarchy of cities in which the larger cities with higher order goods are surrounded by a number of smaller towns as individuals can get lower ordered goods in their own small towns and are willing to travel to the larger cities for higher ordered goods.

There are a series of basic assumptions that underlie the hierarchically laden CPT. First, there is a spatial interdependence among all centers in the model. Unlike Von Thunen's model, no one location exists in "isolation". Secondly, there is a functional wholeness of the system that is made up of discretely stratified places based on size and the associated range and thresholds of commodity supply and demand. Finally, there are catchment areas of goods and services that are provided based on the threshold of the population. The larger urban areas are able to provide much more in the way of higher order goods and services because of the large population and the increased range associated with the limited availability of the commodity or service. The presentation here is limited due to the scope of the book, however, the actual models associated with CPT are much more complicated than we give them credit for here (for further reading, see Marshall 1969). The importance of their introduction here is the fact that the introduction of CPT is an incremental contribution built upon the foundation developed in Von Thunen's zonal land use model, which provided an even broader foundation for the continued development of spatial theories of location.

2.1.3 August Losch: The Economics of Location

While Von Thunen and Christaller's models of economic geography are widely accepted for laying the basic groundwork for what would eventually develop into locational theory in the social sciences, they were primarily concerned with driving force of economics and neglected to look at the role of people or populations. An early examination the interconnected link between location and population was August Losch. Losch's (1937) article, *Population Cycles as a Cause of Business Cycles*, was published in the *Quarterly Journal of Economics*, is one of the earliest attempts to link shifts in economic production and magnitude to population shifts. Losch's approach was different from that of others concerned with the role of population in economics during the same time period as most took the Malthusian approach that population cycles were a consequence of the economy (Losch 1937). On the other hand Losch's thesis was the other way around stating that population is among the main causes of economic changes.

In Losch's *The Economics of Location* (1939), the author identifies the hexagonal isotropic plain as his space of interaction between economics and population. Unless otherwise noted, the following section is in reference to this work. This hexagonal plain is marked by the development of "places" in which individuals, and their aggregates, organize and is developed based on the advanced model of central places in the bottom panel of Fig. 2.2. Losch posited that the hexagonal shape of central places and surrounding peripheral places would be spatially repeating across the isotropic plain with its density and shape tied directly to the good in question. Marshall (1969) notes that the shape of this hexagonal network could be thought of as a fishnet with the mesh of that fishnet directly related to the threshold of the good in question. Ultimately, Losch's idea of central places differed in important ways from the work of Von Thunen and Christaller. Losch believed that the goods themselves created a hexagonal marketplace across space which, when superimposed with all other networks of goods, could be rotated so that centers of sale for goods would line up in the spatial plain. Thus, creating multiple-good supply points. However, because these points are predicated on the maximization of space for the sale of goods, ultimately with an outcome of efficiently driving excess profits to zero, they did not cleanly align with Christaller's CPT. Instead, Losch's model produced a much less centrally structured hierarchy of tiered places and, to that point, is thought of as non-hierarchical (Johnson 1967; Marshall 1969). In essence, Losch's model brought to light an important theoretical contribution concerning the interaction between populations and their environment. To that point, once population is actively involved in the modeling of the organization of places we see a much less neatly organized hierarchy made up unstandardized arrangements of places and distances.

Furthermore, Losch was able to show that this arrangement, specifically in regards to the magnitude and central pull of places, was dynamic and the shifts being felt demographically were directly impacting shifts in economic cycles (Losch 1937). As the principal concern of his 1937 publication, Losch was interested in the role societal advancements were having on the economy, namely the role of longer life expectancies and the increased burdens of the elderly, the increase in birth control and its effect on the capital market, and fluctuations in population increase (1937). Losch saw such population shifts, especially the fast-growing populations of newly developed cities and towns, as effecting business through the large cycles of population movement and migration patterns. To examine this hypothesis, Losch introduced a number of rather sophisticated, for the time period, co-variation tables looking at population cycles and economic cycles. In his final analysis he was able to successfully link the temporally lagged shifts in the development of economic diversity and production, to preceding growth in population aggregates. In support, the simple social fact that business itself locates in areas of population is perhaps the most fundamental evidence of the inter-relationship between populations and their location in space.

2.2 Other Important Contributions in the Development of Early Location Theory

While theories of central places remain important in their laying of the foundation for location theory, other important contributions are also worthy of note here. As mentioned briefly above, the historical use of maps in population based analysis is embedded within a history associated with efforts to understand social and physical issues of their time. For instance, John Snow employed the usage of maps to pinpoint the locations of water pumps and deaths associated with cholera, from Snow was able to visually identify a correlation between proximity to the water supply and higher rates of death. Interesting, only a few years prior, Snow published a pamphlet *On the Mode of Communication of Cholera*, which posited that cholera was spread by contaminated food and water (1949). The production of his map allowed Snow to support those claims when a cholera epidemic broke out in England in the mid-1850s. Ultimately, the production of Snow's cholera map was instrumental in the shutting down of particularly contaminated water pumps, thus containing the epidemic.

Similarly, Charles Booth, also working in London, employed a type of hand made chorpleth map to spatially locate clusters of poverty in the city during the late 1800s. In 1889, Booth published *Labour and the Life of the People*, which was the first of a voluminous set of publications aimed at understanding the various the conditions of those in poverty and ultimately surveying all of London. Booth's interest in the spatial location of poverty was coupled by his driven inquiry to understand more clearly the conditions and life chances experienced by those in poverty. Ultimately, the production of the maps Booth used in his analyses provides early historical evidence of class segregation, poverty concentration, and the urban organization of populations related to social conditions. Even more importantly for this book is the early historical usage of maps as a tool for uncovering and addressing research questions that would otherwise be left hidden.

Finally, we save an interesting familial relationship between location theory and sociology for last. Many accept Max Weber as one of the forefathers of sociological theory as his work is no doubt a core interest to any student while in courses commonly classified as any iteration of social theory within the framework of a sociology department. However, fewer probably also are aware that his younger brother, Alfred Weber, was also an academic. Similar to his older brother Max, Alfred had an interest in sociology but began his career as an economist. In his 1909 publication *Theory of the Location of Industry* (translated to English in 1929), Alfred Weber built upon the economic maximization theories, including theories of central places, in developing a model of minimizing costs associated with transportation, labor, and other financial burdens. Weber's model was primarily dependent on spatial distance as a function of weight of finished products and unfinished materials needed in their production, the costs associated with labor, and the development of critical thresholds of population and industrial competition. Thus, providing support for the influence of the Weber family tree on the development of a geo-sociology.

2.3 Chapter Summary

The purpose of this chapter is to highlight the important historical contributions of location theory in the development of a geographically centered approach to sociological research. The development of locational theory has transcended a number of different disciplines throughout its transition from early models of concentric zones in an *Isolated State* (i.e. Von Thunen's early model), to more complex webs of non-hierarchically structured networks of economic zones (i.e. some of Losch's contribution to Christaller's CPT). The traditional economic theories proved to be useful as the foundation of the use of location in theory. In the next chapter, we will review the Concentric Zonal Hypothesis of the Chicago School of Sociology which later applied these models to theory of the development and organization of the urban city of Chicago. Furthermore, we will continue to these starting points as the book progresses to more recent advancements in spatial analyses in Demography,

Criminology, Regional Science, and many other disciplines. And finally, we will compare this approach to recent theoretical propositions put forth by geographic-postmodernism, which makes it a point that the prior work done in locational theory is no longer applicable to today given the development, technological advancement and ease of long-range communication and transportation. As mentioned earlier, the scope of this book and present chapter cannot fully examine the evolution and context of locational theory. However, these selected historical materials provide a framework for an understanding of foundations from which geo-social approaches to research were developed.

The following chapter examines the next step in this development as we link location theory to early American sociology at the University of Wisconsin's rural sociology program and the Chicago School's focus on urban sociology at the University of Chicago.

Chapter 3
Roots of Space in Sociology: Community Sociology at the Wisconsin and Chicago Schools

The previous chapter introduced some of the early developments of spatial location theory. While the likes of Von Thunen and Christaller laid the groundwork for what would eventually develop into an important spatial theory within the social sciences, they were primarily concerned with *economics* and therefore looked past the role of non-economic social relationships. Around the same time, social scientists began building similar models tying location more directly to population distribution. However, they were implicitly tied to economically 'efficient' models. The earliest documented work directly within the realm of American sociology, was that of Charles Galpin (1915) at the University of Wisconsin who identified distinct spatial arrangements in rural farming communities in Wisconsin. Early on, Galpin's *The Social Anatomy of an Agricultural Community* (1915) identified the importance of places. The 'social anatomy' metaphor was used to emphasize the fact that the spatial context of key social institutions is a major influence on people's lives. His work was certainly not isolated as sociologists in Europe were identifying similar relationships (Friedman 1996). After the publication of *Social Anatomy*, the neighboring University of Chicago was developing its own brand of spatial theory, pioneering field of urban sociology, which in essence was a type of "country come to town."

Perhaps it is not surprising that the roots of spatial theory in sociology can be found among those that studied the rural community. For the community concept itself articulates an inherently geographic element in that it occupies a specific amount of space, it is generally organized around the maximally efficient location of populations and goods, and it provides a relatively standardized enumeration through which individuals can be examined. In fact, Ernest Burgess, of the Chicago School tradition, went so far as to say that both urban and rural communities tend to conditioned by the same forces (1930).

In more recent times, community research remains an integral part of sociology, complete with a specialty journal, City and Community (published by the American Sociological Association) and a related specialized professional association and journal, *Community Development* (Community Development Society). The study of

J.R. Porter and F.M. Howell, *Geographical Sociology: Theoretical Foundations and Methodological Applications in the Sociology of Location*, GeoJournal Library 105, DOI 10.1007/978-94-007-3849-2_3, © Springer Science+Business Media B.V. 2012

community also occupies a significant place in the Rural Sociological Society's activity and journal, *Rural Sociology*.

Community-focused research continues to advance much of the more recent spatial methods and theories. Below we focus on the early developments of social location theory with a specific look at the work of Charles Galpin of the rural sociology program at the University of Wisconsin. By tracking some key bibliometric aspects of the emergence of ideas, we identify what is likely to be a controversial and not formally acknowledged finding: how the work of this early theorist, Charles J. Galpin, served as the intellectual impetus of the urban-centered Chicago School. In regards to the Chicago School, we spend the greatest amount of time on the most influential early works of Robert Park and Ernest Burgess. But we're getting ahead of ourselves due to the dominance of their work! Let us articulate where the genesis of their ideas *and* methods originated as they formally acknowledged in print.

3.1 Galpin's Spatial Focus on Rural Communities

Current disciplinary rhetoric in Rural Sociology argues that "space" is a critical new intellectual frontier for the field. However, the origins of spatial thinking and analysis in Rural Sociology date back at least to the community boundary studies of Galpin (1915). In fact, bibliographically, Galpin's work actually had significant impact on the Chicago School of urban ecology. This little known relationship does not discredit any of the pioneering work that was undertaken by the Chicago School, but instead highlights the potential origin of their work in the rural sociology taking place around the same time at the University of Wisconsin. Perhaps their Midwestern proximity has something to due with the almost simultaneous nature of this thinking. Regardless, the early work in geo-sociology stems directly from community sociology with a focus on both rural and urban communities.

What is often called "community research" in rural sociology can trace its roots back to the work of Galpin's (1915) examination of the agricultural community in Walworth County, WI as an organized space through which everyday interactions were developed, maintained, and institutionalized. Presently, the formal rhetoric in rural sociology emphasizes that "Space Matters" (Tickmayer 2000) and that working with space has always been an interesting focus of rural sociology (Voss 2007). However, the same rhetoric has focused on the resistance of considering space among sociologists regardless of disciplinary specialization (Lobao and Saenz 2002) and the slow adoption of spatial modeling among the same group (Voss 2007). One of us has argued that it is actually the *disjuncture* between theory and methods that is to blame for some of this (Howell 2004), while still others make the point that this is simply a process that in no way should be mistaken for a disregard of space in sociology (Voss 2007).

What is not arguable is that within the realm of a 'geo-sociology' is the fact that fresh challenges lie ahead and must be successfully addressed (Voss 2007). A short-run strategy might be to focus on a few "good" theories of the "middle range"

(Howell 2004). This is perhaps one of the most interesting issues concerning the development of spatial theory and methodological approaches in sociology. As historically relevant spatial theories have been recently outpaced by methodological advancements, we find ourselves faced with tremendous opportunities to better understand and develop spatial thinking in a way that allows us to underscore some of the most central tenants of sociology: the effect of society, it's demographic and social arrangement, political makeup, and many other inherently spatial conditions on the individual living within that society.

In this respect, space is certainly not the final frontier among community sociologists, but in many respects it was one of the earliest in the American tradition of rural sociology. Galpin's (1915) early work on *The Social Anatomy of an Agricultural Community* and related publications had more far-reaching effects than contemporary sociology texts recognize. Galpin's maps spatialized some of the same indicators of community that we conceptualize and struggle with today. For example, in one map, Galpin presented the number of local newspapers received by small locales in the Wisconsin study area.[1] In another, Galpin presents the school attendance zones of Elkhorn, WI using a Point-and-Grid map in order to visualize school attendance zones. He used what is technically now known in the GIS community as a "convex hull" around the furthest students' home locations (e.g., Barber et al. 1996) Among his many other early contributions, Galpin further standardized the operationalization of the farm, the village, the community, and the city in the field of rural sociology (with Campbell and Vogt 1918) and documents the early urban-centric migration patterns through a survey of the movements of farm persons to the city (1930).

Concerning the former, Galpin (1918) understood early on the importance of standardization through the formal operationalization of a number of terms and concepts familiar to rural sociologist, yet often measured in a multitude of ways. Today in the United States, we have such agencies as the Office of Management and Budget (OMB), which in the field of demography standardizes the definitions for all bounded geographic entities (see GARM 1994). Interestingly, the work of Galpin may still be useful, almost a century later, as the one glaring area in which the OMB and GARM continue to leave undefined is rural areas (Porter 2010, 2011). However, today we have access to more exact definitions and spatially bounded coordinates concerning the operationalized meaning and locations of metropolitan areas, counties, states, political districts, school districts, and even less recognized entities like tribal areas (see U.S. Bureau of the Census). Such a standardization has greatly improved research in these areas as it has allowed researchers to quickly understand the exact meaning of what area of space is being examined and how that might relate to past research of a similar scope. But Galpin as saw the need for such standardization in the early days of rural sociology, unfortunately rural America remains the least standardized entity by government protocols since rural is basically what is left over after all other areas are removed from the full geographic landscape.

---

[1] As we will show later in this chapter, the importance of this map was instrumental in the development of much of the research of the early Chicago School sociologists, including the work of Robert Park who made many of these methods standard in the discipline.

Among Galpin's other work, the early identification of the push-and-pull factors associated with the migration of rural persons to the city was documented in an article in the *American Journal of Sociology* in 1930. This was by far the dominant outlet for leading sociological research and remains one of the top two journals in the field. It is highly significant that *AJS* was founded and published by the University of Chicago. Galpin was concerned with all parts of rural life. Here he was specifically interested in the population mobility that was underlying the massive movement of rural persons to the city. Over the decade of the 1920s, Galpin believe that it would not be an exaggeration to believe that every farm family had at least one person in the city to which they held a close relationship. He identified this movement of rural to urban migration as unusual, given past population trends, and was quick to identify strong reasons to stay or move that were related to economics, socio-cultural values, and proximities. Ultimately, the movement of populations across space over time continues to be a frontier to which spatial analyses has yet to fully understand.

Concerning the focus of this book, the most historically relevant piece of published research by Galpin is the 1915 *The Social Anatomy of an Agricultural Community*. This research was based upon a social survey of Walworth County, WI that used as its primary tool, large working maps of the county. According to the research bulletin, large maps were created, with a scale of 2 in. to a mile, and thumbtacked to a large board. From there they were later transferred to paper in the format that they are presented in the bulletin. Within the county, Galpin selected 12 villages and cities to be surveyed in each of which he selected informants with knowledge of the local community and social networks associated with its trade patterns. Galpin surveyed teachers, bankers, clergymen, librarians, and others in an attempt to understand spatial connectivity issues associated with the farthest travel a farmer would make to trade in the village from all directions, which would result in a rough trade boundary. Next, bankers were surveyed in each of the 12 communities in order to identify long-distance farmers that bank in the town. Further information was gathered on school district inclusion and newspaper circulation. As a result, Galpin identified a series of "communities". He created separate maps in the research bulletin related to the "trade communities", "banking communities", "newspaper communities", "milk communities", "church communities", "high school communities", and "library communities". Ultimately, Galpin concludes that it is trade zones that form the actual boundaries of these communities, in which the lives individuals make up the larger unit of the community.

His work does invoke a very intriguing graphic representation of central places and their overlapping fringes, which harkens back to the work of the Von Thunen and Christaller economic models. In this regard, Galpin points out that those living on the fringes of communities are likely to the part of the social anatomy of multiple communities in regards to their day to day lives. However, of the greatest import is the use exemplary usage of isotropic maps in the identification of community. Similar maps were later developed by other rural sociologists, including dot density maps of Frey and Smith (1936), which used these methods to visualize data from the 1935 Census of Agriculture.

The impact of Galpin's work in early American urban sociology was tremendous, but is actually rarely recognized today. In a 1929a, article in the *American Journal of Sociology*, Robert Park mentions that

> some thirteen years ago, Dr. C.J. Galpin published…a stimulating little paper called *The Social Anatomy of an Agricultural Community…*The purpose of the paper was eminently practical… Its method was to plot, on a series of maps, the actual relations—economic, political, and social—of the farm populations… These observations have given rise to the natural area… Since 1915, further studies of the kind that Galpin initiated have been made of urban communities…, the same methods have been extended… to the city, its suburbs, and to the work regions in which the city dominates.

This direct attribution of Galpin's work as an influence on one of the most prominent schools of thought in American Sociology, the Chicago School, is rarely recognized. Although it is fair to note that Galpin himself may have benefited from the early work of Albion Small and founders of the first Department of Sociology at the University of Chicago whom routinely used maps a way of illustrating social relationships. However, the importance and influence of Galpin's introduction of the isotropic map as a method for understanding social organization is largely understated in the history of the discipline and urban sociology. In fact, the Chicago School's land mark work (*The City*), was published without an acknowledgement to the intellectual debt owed to Galpin as he was not even cited. In a sort of domino-effect, similar slights continue to exist. For instance in a book describing the first century of the Department of Sociology at Chicago, Galpin is also not cited even though his bibliographic impact was tremendous (see Abbott 1999). To many this is somewhat understandable; it was actually the Chicago School that took these seminal ideas of the so-called "natural area" and the nascent methods of spatial visualization to a new level. Yet it remains clear, per Park's own statement above, that the innovative approaches developed by Galpin provided a framework from which some of the most influential work of the Chicago School was put forth. Even today techniques from Galpin's early work continue to be useful in geo-social research. For instance, a technique now referred to as using a "convex hull" around outer-most points articulating some characteristics—in Galpin's case, travel to identify community boundaries—is now used in many GIS studies of behavioral phenomena (Barber et al. 1996).

In the same vein, other research from rural sociologists was also being published at the *American Journal of Sociology*, which likely provided a strong form of academic discourse through which many of the ideas of what is the Chicago School first developed. It is important to note that during this era, rural sociologists were viewed as having some of the strongest "field research" (no pun intended) programs in the United States. Part of this was no doubt due to the Department of Agriculture's funding of social research on rural communities (see, for example, Sorokin et al. 1930). Some of these include the comparative focus of rural and urban social organization by Terpenning (1928), where the primary scope was on the capacities of the communities to provide for its residents with a direct focus on the inadequacies of the needs of rural residents were being met. Ultimately, Terpenning suggests a tentative program for meeting such needs of rural residents given the vast different conditions through which social organization and group consciousness occur in rural and urban communities.

This focus on the community in the early work of rural sociology led it to be largely considered the leading specialty in Sociology in terms of how social science work should be undertaken. For example, this argument was made by Bruce Melvin (1927) in the *American Journal of Sociology.* It revolved around the methodological advancements in the discipline that were moving sociology beyond the philosophical and towards the scientifically methodological. He goes on to make an interesting comparison concerning the trajectory of Biology as a discipline following the advent of the microscope. In a similar way, Sociology was developing methodological tools at that time that were moving it more and more in the direction of physical science.

One of the most influential methods of that time in Sociology was the advent of methods aimed at identifying and measuring social organization of population through the usage of isotropic maps (Mowrer 1938). This method was used early on by Galpin (1915) and then made well-known over a number of decades by the continually innovative approaches to understanding space and the organization of humans in space by many researchers associated with the Chicago School. For example, Mowrer (1938) writes that Galpin was able to differentiate boundaries of "natural areas," delineating one rural community from another. As Park noted, using this method Chicago School researchers were able to discover 75 different localities within the City of Chicago. Mowrer's maps, as with much of the work at the Chicago School during this time, were explicitly interested in mapping social behaviors and natural areas in which social pressures (clearly an ecologically-rooted concept) could be assumed to have been similar on the population occupying those areas. Furthermore, in the sense that Galpin's method was both practical and perhaps even somewhat simplistic in its delineation of rural communities, the Chicago school found much more diversity and heterogeneity in the metropolis. This ultimately led to a much more dynamic application of mapping and spatially-guided analysis into issues that had always been at the heart of sociology at-large, social problems.[2]

3.2 Spatial Theory and Location in the Urban City at the Chicago School

The use of location can be seen as a trademark of a number of researchers associated with Chicago School sociology, in that, much of their work was focused on the city of Chicago and was based on spatial zones throughout the city, which helped to

[2] An interesting note on this matter is that is what rural sociologists had funded by the U.S. Department of Agriculture. The improvement of "country life" was at the core of research plans published routinely in the *American Journal of Sociology* during that era (Melvin 1927). It is a far cry from the editorial visions of most of the top journals in the field today.

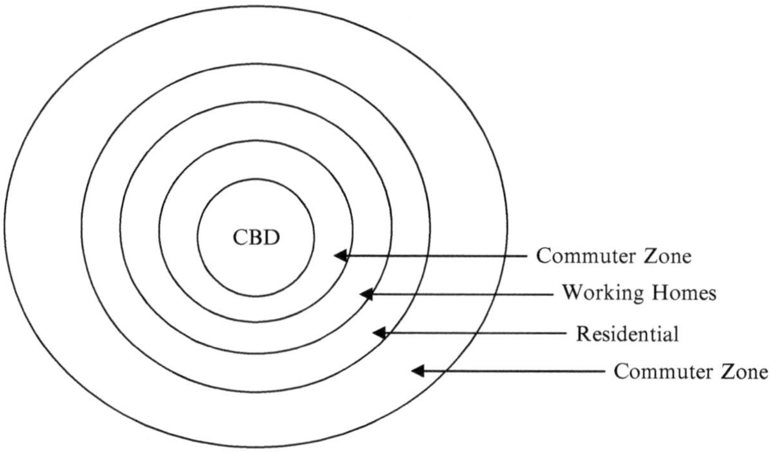

Fig. 3.1 Park and Burgess' concentric zonal model of urban ecology

characterize a "typical" person based on predicted attributes. *The City* (Park and Burgess 1925) introduced an idea that would become known as the concentric zonal hypothesis, following the earlier work put forth by Von Thunen, Christaller, and Galpin, (and among others, no doubt). The Park and Burgess (1925) hypothesis posited that the city of Chicago could be seen as having a central core with rings dispersing outward, much like the Von Thunen economic model. The rings, then, would indicate particular zones that could be labeled based on their characteristics. The original hypothesis had five zones: (1) Central Business District, (2) Transition Zone, (3) Workingman Zone, (4) Residential Zone, and (5) Commuter Zone (Park and Burgess 1925). Each zone can be seen as pushing outwards into the next zone from the Central Business District. The zonal model is presented in Fig. 3.1.

The development of this zonal model, through an elaboration on the preceding spatial theories and methods, created a distinctive ecological research program at the University of Chicago that became known as the Chicago School of Sociology. The research maintained an interest in the development and maintenance of the city as an ecological environment, complete with the same evolutionary models associated with nature and a specific focus on the ability to secure resources. Park and Burgess believed that one of the most important resources in the city was land and that this land would be prized at different social and monetary values. Ultimately, this would result in the partitioning of land into "natural areas" whereby people would be selectively organized into given similar social characteristics. These natural areas would make up the concentric zonal model presented above.

Other work by Robert Park directly drew more directly on the social organization work of Galpin in the measurement of urbanization measured by newspaper circulation (1929a). In what Park described as an "eminently practical" approach to social research, he went on to apply the same method Galpin used in Walworth County,

WI to determine the extent of urbanization in the city of Chicago by newspaper distributions, given the assumption that culture is based on communication. What Park finds is an interesting distribution of urbanization that not only highlights the actual city of Chicago, but also the periphery areas that enjoyed high levels of newspaper circulation during the time.

The maps in Park's 1929a article were some of the earliest to also underscore one of the more stable conditions of community development and urbanization: the undeniable relationship to between communication and transportation. What Park found was that the peripheral areas of "urbanization" were spread along railways and highways and included the annexation of towns and villages that had once existed as independent entities. Thus, Park was able to identify one of the binding factors still used today by the U.S. Office of Management and Budget in the delineation of metropolitan areas, economic and social integration with the urban core based on transportation and communication linkages. Of course today, with advances in telecommunications and transportation, the ubiquitous slogan in real estate of "location, location, location" may seem to have lost its edge (see Friedman 2006), but critics will quickly point to increasing rates of residential segregation, migration patterns, and geographic variations in any number of indicators of human well-being as a counter.

In Park's 1929a article, he also uncovered a relationship specific only to large metropolitan areas. This relationship harkens back to the hierarchical structuring of places that was introduced in purely economic terms by Christaller and Losch, but it identified socially linked satellite towns and villages to the metropolis of Chicago. It seemed that much in the same way that lower-ordered goods were available at differing rates based on the size and range of the community, so too were the social connections of places to the cultural influences of news from higher-ordered places. Park showed that distinct satellite regions could be delineated when newspaper circulation was taken into account. These satellites are familiar to anyone who is familiar with the Chicago area. He identified areas like Waukegan, Evanston, Hammond, Gary, Elgin, Aurora, and Joliet as satellite population centers that would serve even lower-ordered towns and villages proximally situated to Chicago.

As such, Park was able to ascertain that over half of the newspaper subscriptions in these satellite areas were circulation from Chicago. Park goes on to call this the "devolution" of the metropolis and what could be considered the development of and changing character of small towns given the increased usage of the automobile. In essence, this was the first formal identification of what would later become known as population deconcentration, or the movement of individuals to the suburbs and rural areas given the ease of transportation and communication (see Lichter and Fuguitt 1982, 1989). As a result, Park makes a point to note that as cities had become independent of rivers and railways, given the increasing usage of the automobile, they took on a more symmetrical and radial form as the urban center stretches outward. This indicates that Park saw the concentric zonal model to be more applicable to the developing city of that day given the loss of restrictions to growth based on proximity to rail and water.

3.3 Chapter Summary

Some of the earliest work in spatial thinking associated with the development of American sociology as been highlighted in this chapter. Along with the development of the discipline in the U.S., it is clear that the sociology departments of the Midwest dominated this early way of thinking and the development of associated methods. There were also similar movements in spatial thinking were taking place around the world. In France, for instance, Marc Bloch was calling for an interdisciplinary approach to the understanding of social phenomena (Friedman 1996). At the core of such a cross-discipline approach was the integration of sociology and geography. Friedman (1996) points out that Bloch advocated an approach somewhere between sociology and geography, focusing on the sociological impacts of concepts like natural boundaries using cartographic depictions and how they facilitated the understanding of an ecologically distinctive social-system. As such, the early 1900s provided some of the most influential theoretical and methodological advancements the creation of a geo-sociology. Most importantly, we will see in the next chapter that as theoretical progress continued to be made by the likes of Amos Hawley and other human ecologists, the same advancements were not taking place in regards to the methods used in analyzing such relationships in space. As a result, we argue that many in the field moved away from ecological analyses and towards more individual analysis of social behaviors.

Today, we have the *opposite* problem. As methods in the analysis of individuals in space have continued to advance, we find that the decided lull in theoretical development created by the discipline's movement away from ecological analysis has left the modern researcher with a large choice of methods for spatial analysis, but very little in the way of theoretical orientations to guide such research (Howell 2004). While the mapping techniques of Galpin, Park, Burgess, and others could be considered much less sophisticated using today's GIS procedures, they did have a large dose of theory that directly guided the development of the concepts of "natural areas" and "concentric zonal models".

It is also undeniable that the foundations set by the rural sociology program at the University of Wisconsin and the Chicago School of urban sociology are immensely valuable in their contribution to spatial thinking. This is true for the former regarding the largely unacknowledged but exemplary work of Charles Galpin and the later for the tremendous impact it has had on the discipline of Sociology as a whole. Later in this book, this impact will become increasingly evident. In this chapter, we have spent a little more time discussion the contributions of Galpin and other rural sociologists for two reasons: (1) they are less well-known among sociologists and those whose work takes a geo-sociological approach to there research and (2) because much more is available on the work Chicago School, including later in this book.

The next chapter focuses on the continued development of human ecology in the mid 1900s. Our primary focus will be on the work of Amos Hawley and his famous theoretical essay (1950), which is still considered one of the most influential sources in the revival of the human ecological perspective and macro-sociology in general. The subsequent chapter will lay the foundation for theoretical advancements in geo-sociology that have recently been made or that are likely to be made in the future.

Chapter 4
Human Ecology and Its Link to Geographical Sociology

4.1 Introduction

Human ecology is largely interested in the social organization of populations and the effects of environment and larger social structure on individuals within a given macro context. At the turn of the twenty-first century, sociologists have become enamored with the effects of the environment through applications of statistical methods take *place* into account as a factor in variations in human behaviors and social processes. However, during the early 1900s these methods were far from being advanced and the primary approaches to spatial analysis *in sociology* consisted of the "hand-produced" methods used by Charles Galpin and the early Chicago School as noted in the previous chapter. The next half century saw an unprecedented boom in spatial thinking that has continued to provide the framework—sometimes without explicit recognition—for much of the spatial analysis that is currently undertaken today.

It is at this intersection of place, culture, and social process that the actual placement of human ecology has been argued over the years. This is especially true in the early twentieth century when the state of sociology as an academic discipline was still very much in development. However, the early focus of some of the most influential American sociologists at the Chicago School on the relationship between man and his environment, made the boundaries among ecology, geography, and sociology even less clear. Some argued that sociology was part of a general discipline of ecology (Adams 1935) and others argued that geography was itself to be considered part of "human ecology" (Moore 1920). Ultimately, this type of discourse was rather prominent and underlies the inter-disciplinary development of human ecology from its very beginnings.

At various points, the search for niches within human ecology caused some scientists to try to understand the unique roles of sociology and geography within the newly developing approach. Robert Park most prominently drew the line between the two by focusing on the core interests of the two disciplines (Gross 2004). Park (1926: pp 487) stated that "Sociology starts with society, but geography starts with the soil" and further notes that "Sociology seeks to classify its facts and to describe social changes in terms of processes". Thus, sociologists found interest

J.R. Porter and F.M. Howell, *Geographical Sociology: Theoretical Foundations and Methodological Applications in the Sociology of Location*, GeoJournal Library 105, DOI 10.1007/978-94-007-3849-2_4, in trends, patterns, and "social facts" whereas geographers were much more interested in the irregular or actual (as opposed to the typical).

In the mid-1900s, the role of space and the ecological environment experienced somewhat of a boom in Sociology with the development and refinement of a human ecological approach to studying populations. Perhaps the most famous text linked to this development was Amos Hawley's, who's *Human Ecology: A Theory of Community Structure* (1950) and *Human Ecology: A Theoretical Essay* are still the considered by many to be (if not *the*) primary reads in the area of modern human ecology. Many other important works were also developed during this time, all of which were greatly influenced by the most seminal work in this area which was discussed by Emile Durkheim as "Social Morphology" (earliest known published materials on the subject appeared from 1897 to 1898), and the formalization of the term human ecology in 1921, by Robert Park and Ernest Burgess in their edited *Introduction to the Science of Society* (Schnore 1958). However, this formalization of the term was not actually the first mention of "human ecology" as a term for the study of *Geographic Sociology*. Instead, Edward Hayes' 1908 article in the *American Journal of Sociology*, "Sociology and Psychology, Sociology and Geography", takes that distinction (Gross 2004). Hayes' linkage to the Chicago School is also direct as he was a student of the founding Department Head at the University of Chicago's Sociology Program, Albion Small. As such, the preface of Hawley's essay mentions that those to whom credit is owed is to numerous to begin listing, however, special acknowledgements were given to Park, Burgess, and Mckenzie, as well as the "Chicago School" itself.

Hawley built upon the work of others in order to develop the ecological foundation in which the development and maintenance of the community takes place by putting forth a series of empirically grounded assumptions and hypotheses involving the intersection of the natural-environment and population phenomena. Conceptually, the piece has been essential to the development of spatially-centered empirical analyses and continues to compliment advancing spatial methodologies. As of the time of its publication, these methodologies continued to lag behind the tremendous amount of ecological theory that was being developed at this time. For instance, the foundational work supplied by Galpin, Park, and Burgess, and other early sociologists of community studies provided the framework from which theoreticians could begin to see communities as important determinants in human behavior, and conversely, as directly shaped by human nature. However, the lack of available technology kept the methods somewhat simplistic at this time, consisting largely of isotropic maps and trend analyses. As theory began to push the methods of ecological analyses forward, we begin to see the borrowing of established methods of plant and animal ecology being employed in the field of human ecology.[1]

[1] This is not the first time that what became a significant analytical procedure from another discipline was imported and used successfully in sociology. For instance, Othis Dudley Duncan's adoption of "path analysis" from genetics for use in the land mark *The American Occupational Structure* (Blau and Duncan 1967) is a well-known example (see also Duncan 1966). For other examples, one must not search too long in the journal *Sociological Methodology*, which is littered with cross-disciplinary applications.

Early advances in such methodologies began to take place in the 1960s as outlined by Sorokin (1964). Here, Sorokin points out such advances as driving an accelerated understanding of spatio-temporal causality. However, he is quick to acknowledge that, while such relationships have their founding in the natural science, the application of such, under the same methodology and terminology, would be irresponsible as social relationships occur at a much different level. Thus, a human ecological approach was one borne out of the work of early plant and animal ecology. Specifically, the ideas put forth by Durkheim, and his theory of "Social Morphology," are said to be directly influenced by the work of plant ecologist Ernst Haeckel who coined the word "ecology" in 1868 and the work of Park and Burgess directly attributed the concept of human ecology to social morphology (see Schnore 1958).

The following sections of this chapter will guide readers through the historical development of the human ecology approach to social research. Of course the scope of the chapter will not allow for an exhaustive account of all of the influential literature in this area, but does highlight the continued development of a geo-sociological approach grounded in the further development of location theory and human ecology. As such, we follow this section with an expanded account of the information presented in this introduction; including Durkheim's early interest in the social organization and population phenomena, Edward Hayes' first publication and Robert Park's formalization of the term 'human ecology', Hawley's theoretical essay, and other important developments leading to the modern state of spatial thinking and methods in sociology.

4.2 Durkheim: Sewing the Seeds of Human Ecology

In understanding the roots of human ecology one must understand that "ecology is commonly defined as the study of the relation of organisms to their environment" (Hawley 1950: pp. 1) and relationally "the 'web of life,' in which all organisms, plants and animals alike, are bound together in a vast system of interlinked and interdependent lives…" (Park 1936a: pp. 1). The term ecology was coined by Haeckel, a plant ecologist, in 1868. Haeckel's usage of the word was further in direct reference to the work of George Perkins Marsh, whose 1864 book entitled *Man and Nature* served as an early argument against deforestation. Marsh argued that the deforestation of areas of the U.S. would ultimately lead to the development of large desert regions using examples of historical forest clearings and subsequent consequences. In explaining such relationship between man and nature, Marsh developed the phrase *the economy of nature*. In direct reference to this phrase, Haeckel introduced the term ecology (Odum and Barrett 2005). It is from these roots, that the terms made its way into the field of sociology.

Schnore (1958) brings to light this relationship in his article "Social Morphology and Human Ecology" in the *American Journal of Sociology*. Schnore references the influence of Durkheim's concepts of Social Morphology as being relatively small in regards to his impact on the discipline of sociology, but notes that the concept was to be devoted to two primary interests; the first being the relationship between the environment and social organization and the second being the operational inquiry into population size, density, and distribution. As such, Durkheim's Social Morphology sewed the seeds, within the field of sociology, for what would become human ecology. Furthermore, Schnore highlights the fact that Durkheim was fully aware to the structural conditions of interdependence that must be present for any complex system of organisms to exist. However, the ecological view was something that did not appear in Durkheim's writing until relatively late in his life (see *Sociologie et Sciences Soiales* 1909 for example).

The first usage of the term "human ecology" in the sociological literature is attributed to Edward Hayes, following the publication of his article in the *American Journal of Sociology*. At the time, Hayes reminded readers that "uncertain the status of sociology as a specific science as yet may be, a movement of scientific inquiry into the social field of, from various sides, … bent on contributing toward the explanation of the social reality" (1908: pp. 387). As such, Hayes points to existing subfields of "social geography" in his exclamation that such a sub-field was fledging in the developing discipline of sociology. In fact, Hayes notes that "geographic sociology" is concerned with the physical conditions that determine social phenomena. He mentions climate, natural resources, mountain barriers, rivers, harbors, etc., as being similar to the geographic features which sociologists should concern themselves in the examination of population processes (pp. 387). Of course, this early identification of a geographic sociology, what we now refer to as geo-sociology, is overwhelmingly interested in the role of the geographic in the physical sense of the landscape. Today, we see that geo-sociology continues this interest, but also makes its interest in the proximity of populations to other social phenomena.

In the 1921 book, *Introduction to the Science of Society*, edited by Park and Burgess, the term human ecology reappears for the first time in a sociology textbook. It is at this point that the term is formalized as a meaningful concept to be discussed in courses in the discipline of sociology. Thus, much like the augmentation of Galpin's early maps, Park took the ideas put forth by Durkheim, Hayes, and other ecologists and made them meaningful to the early development of American sociology. Chicago School predecessors of the Park and Burgess team also figure quite heavily into the line of thinking associated with human ecology. For instance, the first American textbook for sociology, *An Introduction to the Study of Society*, was written by Albion Small and George Vincent in 1894 (Gross 2004). While the two never used the term ecology directly in their monograph, the focus on geographic and ecological methods was nonetheless a central focus of the textbook. In particular, the study of the interactions between the social world and nature were often depicted in ways that would be familiar to ecologist and geographers of the time, including the usage of methods aimed at understanding the flight patterns of birds in order to understand the mobility of individuals within the city (Gross 2004).

It is evident then that the earliest roots of human ecology within American sociology can be traced to the Chicago School and the inception of its department under Albion Small and colleagues. In fact, the earliest ideas and frameworks for the study of sociology were presented in textbooks written by those in the department whose primary interest was the looking at the world as a natural laboratory. The approach included the geographical approaches to understanding population distributions, size, density, and so forth. The influence of the Chicago School carried on to the actual first publication of the term "human ecology" in the 1908 *American Journal of Sociology* article in which the former University of Chicago student, and head of the University of Illinois Sociology Department at that time, quoted private communications with J. Paul Goode (Gross 2004). Goode was a geographer in the Department of Geography at the University of Chicago during this time and is the namesake for the *Goode's World Atlas* (Gross 2004). The interdisciplinary nature by which the term "human ecology" was developed and presented as an academic interest highlights the route by which human ecology aimed to understand social phenomena through an understanding of the geographic conditions in which the individual was situated. Following the work of Albion and Small, Hayes, Goode, and others the Chicago School remained prominent in the development of a geographic sociology through the vast amount of literature contributed on subject of human ecology and urban community studies by Robert Park.

4.3 Robert Park's Human Ecology

Robert Park wrote extensively on the topic of human ecology. However, in doing so Park was quite interested in separating the role of sociology from that of other similar disciplines, such as geography. In this sense, Park's 1936 article (Park 1936a), aptly titled "Human Ecology," aims to make two things implicitly clear; (1) ecology in all senses—plant, animal, or human—has central to its primary interest the study of 'community' and (2) there are distinct features among each of the three ecological approaches that necessitates a distinct approach in their understanding. Core among these indicators was the role of the non-biological variant culture. Culture and its artifacts, according to Park (1936a), consisted of any number of social variables that a biologically-centered ecological viewpoint could not account. Park points out that "the cultural community develops in similar ways to that of the biotic, but the process is more complicated" (pp. 9) and that "inventions, as well as sudden or catastrophic changes, seem to play a more important part in bringing about serial changes in the cultural… community" (pp. 9). In essence, Park ensures that the role of sociology in human ecology is not seen as a sub-discipline of Biology or Geography, but instead as an independent field of scientific inquiry to which the core interest is human ecology.

Park (1926) further points out that human ecology is not the same as human geography in that it is not the individual that human ecology finds of interest. Instead, it is the *community*. The actual social organization of man in space as opposed to the relationship of man to geographic proximity is essentially of primary interest. Park (1926) states that an individual's "relations to other men, that concerns us most" (pp. 3). Thus, the social organization of space, in contrast to geography, is of the most interest to human ecologists and, as proof, the identification of typical patterns and natural areas dominated much of Parks early interest in human ecology.

Park (1926) indicates that of specific interests within this realm are the relationships of communities to one another, population distributions, densities, and patterns of segregation in the largest of cities. Not surprisingly, segregation was the primary focus of much of the early Chicago School work on natural areas in that it focused on the separation of individuals into areas of the city by class, race, and occupation. Furthermore, this decidedly human focus of Park's human ecology was intentionally designed to drive a distinction between the study of human communities and the plant and animal communities studied by the existing natural ecologists.

This does not mean that Park abandoned the thinking of ecologists. Instead, Park simply took many of the terms well known to ecologist of the time and applied them to human civilization. In many cases, the primary focus of Parks human ecological focus was the urban center or city. In doing so, Park introduces ecological terms such as 'competition', 'dominance', and 'succession' in a manner that is directly applicable to the study of aggregate populations of humans (see Park 1936a: pp. 51 for instance). According to Park, in the same way that dominant plants and animals were able to 'win' in their competition for valuable resources, human beings also compete for valuable resources. In previous work, Park (1934) related the idea of dominance to the natural landscape and the approach that a geographer might take in noting that the highest mountain in a region dominated the landscape. This dominance not only had to do with the sheer size of the mountain, but also its likely impact on the winds, plant and animal life, and climatic characteristics of the region.

At this point, Park (1936a) directed his attention to a very familiar idea to sociologists; the zonal areas of metropolitan communities (in reference to his work on *The City* 1925). Park explains that the metropolitan area is made up of many different zones; slums, shopping districts, business districts, banking centers, etc. In this line of thinking, each of these zones exist, and owes their existence, in accordance with the factor of dominance and indirect competition for the most valuable space in the area. Thus, patterns of accessibility and financial resources are most directly related to the demand for space and ability to attain that at high land value (Park 1936a). Ultimately, the dominance of some institutions and groups in the human community can be linked to the succession of others. Furthermore, Park (1934) notes that dominance can be seen as a relegation of individuals and races to specific niches through which they are likely to face the least amount of competition and be able to maximally contribute to the rest of the community. Thus, dominance is the medium by which the macro level competition of a society shakes out the social hierarchy and organization of the various groups in the community. As a fluid state, the equilibrium that is achieved through the dominance of one group or class within society is often overcome through social processes of change. In these cases the natural change, or what Park (1936a) refers to as succession, begin to shape the organization of the community.

Succession, in Park's view (1936b), is the process by which a community moves through a series of relatively ordered steps. It is at this point that Park points out the relatively high level of susceptibility that the human system encounters in relation to the plant and animal systems. This makes the human system distinct and to some degree more dynamic and less fluid, than the direct application of a biotic approach to the human community many suggest. Park (1934) seems to indicate that the social contract through which the communal nature of the human community was created provides a state of equilibrium through which the individual, within the community, is bounded by a series of laws, norms, values, and other expectations of behaviors. These personal relationships to the community further restrict personal freedom, but allow for the development of an interdependent nature of collaboration and consensus. As the actual physical shape, size, and density of the human community changes over time, through processes such as expansion and contraction, these personal linkages of the individual to the community social contract shift as well. Such shifts indicate a succession, or stable change, in the social order of the human community that directly affects the relationship of the individual to community as well as man to man. Such instances of succession in the plant and animal community are much more likely to be related to the introduction of new species, changes in light/water/nutrition availability, and other natural occurrences.

As an example of this distinction, Park points out human beings are not as dependent on the environment as animals. Most central to this is the idea that humans can overcome much of their environmental dependence given their shifting reliance on other men (Park 1936a). Here Park points to an essential difference between the human community and the animal communities, the ability to divide tasks. The division of labor has allowed man to specialize and create interdependent networks through which he performs a narrow series of tasks in return for providing services to others who perform unrelated tasks and provide resources in a reciprocal manner. Yet the lessening of man's dependence on his local environment is not to say that man is freed from ecological determinants of social outcomes. In fact, Park seems to suggest that the communal nature by which humans have removed the "chains" of locality serve to impose its own set of regulations on the human system. At this point, man's reliance on nature is replaced by communally agreed upon culture, customs, and the development of institutions (Park 1936a). All of which vary from society to society and serve to, both directly and indirectly, influence the behaviors of the individual and the aggregate population.

Thus, "human society, as distinguished from the plant and animal society, is organized on two levels, the biotic and the cultural" (Park 1936a: pp. 14). This is an essentially important point to Park. To this point the foundation of human society as a distinguishable entity from that of other types of communities has been the primary focus. However, at this point it is the theoretical importance of consensus and social organization that begins to reveal itself in Park's theory. In that respect, Park believes that competition is unchecked in the animal and plant kingdoms, but the very ability of humans to divide labor and establish interdependent relationships based on consensus and cooperation, theoretically diminishes much of the inherent competition in the biotic state. Most important to this belief is the fact that freedom to compete in the human community is directly restricted by moral, political, and economic conventions, laws, and understandings (Park 1936a). Thus competition is still evident in the human community but greatly restrained in comparison to the plant and animal communities to which the original understandings of ecology were developed.

However, the role of dominance and succession remain a formal part of Park's human ecology. He concerns himself with the dominance of individual groups and institutions which create a "biotic balance and social equilibrium" (1936a: pp. 15) that must be investigated in regards to two primary interests; (1) the establishment and maintenance of such levels of balance and equilibrium and (2) the stable transition from one stage of balance and equilibrium to the next. Thus, the first interest of investigation concerns the dominant mode of social organization and the second the succession of states through which the human community changes over time and in relation to a series of components that make up the community. Park (1936a) introduces these components as (1) population, (2) material culture, (3) non-material culture, and (4) the natural environment and highlights the core interest of human ecology as an interest in the physical and cultural movement of each as well as changes in the relationship between man and the natural environment.

Perhaps most interesting in the development of Park's theory of human ecology is his lack of interest in referencing biologists, geographers, and other ecologists. Perhaps Park's drive to develop a sociological dominance within the field is to blame, or perhaps a sincere belief that the previous understandings of plant and animal ecology could not contribute to a distinctly different viewpoint on the social organization of humans. In doing so, Park essentially "rediscovered the wheel" through the introduction of terms and theoretical ideas presented by those close to, and in some cases part of, the Chicago School only a few decades earlier (Abbott 2001; Gross 2004). However, the efforts put forth by Park to establish a sociological focus as the main emphasis of human ecology and the extensive writings developing the theoretical framework supporting that distinction provided a framework from which studies of human ecology drew for years to come.

In short, the primary goal Park's human ecology is that it advocated for the establishment of the *community* as the primary unit of analysis in sociology. Park (1929b) saw the community as "the habitat in which alone societies grow up" (pp. 7). This along with the fact that communities are visible entities, complete with physical boundaries and specific geographic locations for institutions, make the community more appropriate for statistical treatment (Park 1929a). In making these types of arguments, Park essentially identifies the community as the primary focus for sociological research.

Today we understand that much is to be learned from an understanding of relationships of individuals to social institutions. However, Park's point is that each of these individuals exists within a community, which when not taken into account may leave a large amount of variation unexplained in any outcome of interest to social scientists. To put it more clearly, Park (1939) makes the statement that even to the most disinterested observer, "society is obviously a collection of individual living together, like plants and animals, within the limits of a common habitat, and it is, of course, something more" (pp. 1).

This collection of individuals into communities were to be the focus of human ecology in a way that would allow social scientists to standardize their research in regards to a common unit of measurement and in regards to an agreed upon set of interests. Core to these interests were the actual changes in the physical and cultural community, in terms of size, organization, demographic makeup, cultural beliefs, etc. that would come to represent stages of succession driven by shifts in dominance and ultimately manifesting itself as measurable changes in the community. Such ideas as expansion and contraction come to mind when one thinks of the physical changes in the community and its borders. Other concepts such as mode of production transitions and cultural shifts due to immigration and the influx of diverse populations also come to mind. These ideas remained strong in the development of human ecology through the middle of the twentieth century and provided the framework form which other influential work drew its inspiration.

The next section introduces on such influential human ecologist through the introduction of the work of Amos Hawley.

4.4 Amos Hawley and the Evolution of Human Ecology in Sociology

At the time of Amos Hawley's theoretical essay, *Human Ecology: A Theory of Community Structure*, many researchers within the fields of economics, psychology, and medicine had taken to research which they referred to as human ecology. As Gross (2004) points out, besides the name, they had very little in common with the concepts and interests that were at the core of human ecology in the sociological sense. However, the work of Park and the Chicago School continued to remain influential in the research of many sociologists. Perhaps the most historically prominent writing in this area during this time period was the before mentioned essay by Amos Hawley. In fact, in 2009 Donald Bogue, a student of Hawley, wrote that "[Hawley's] writings rank with those of Weber and Durkheim in explaining social organization and excel then in explaining social change" (pp. 1). Furthermore, Bogue went so far as to name Hawley as the father of environmental effects.

Today the sociology literature is replete with studies and papers interested in the effect of neighborhoods, communities, and the environment; if we are to believe the words of Bogue then much of this work is indebted to the foundation laid by Amos Hawley. W. Parker Frisbie (2009) makes similar statements concerning the place of Hawley in the discipline. Frisbie says that Hawley's scholarship has no doubt secured him a place among the giants of sociology, and this would be true if his only contribution had been his 1950 *Human Ecology*. His influence is believed to have broadened our understanding of macro processes of population, organization and development (Hiday 2009), been a major influence on the evolution of twentieth century social theory (Hirshman 2009), and even defined the field of sociological human ecology (Polston 2009).

While these comments refer to much of the published work of Hawley, the acknowledgements tend to have general slant towards to relative importance of *Human Ecology* to the field of sociology and the sub-field of human ecology. Here we will spend time highlighting the most important points from the 1986 essay *Human Ecology: A Theoretical Essay*, which build directly off of much of Hawley's work preceding its publication and provides a succinct presentation of Hawley's human ecological theory. Most notably, the essay builds upon Hawley's 1950 *Human Ecology: A Theory of Community Structure*, for which he is perhaps most famous.

Hawley's human ecology builds specifically on the work of Park, Burgess, and McKenzie. In fact, the first page of chapter one in *Human Ecology* specifically references their work concerning their treatment of cities in much of the research they published in the 1920s and 1930s. Hawley (1950, 1986) sees human ecology as an attempt to deal with social organization in a holistic manner, but maintains that the ecological perspective alone is not enough. The ecological perspective must be informed by the sociological perspective. So just as Park believe that the human ecology must include the components of culture in order to understand the distinctive traits of human communities, when compared to plant and animal communities, so to did Hawley believe that a purely ecological perspective was flawed in its inability to account for that very same distinctiveness. As Thomas (1992) points out, Hawley's "new ecology" built upon the groundwork laid by Park, Burgess, and others at the Chicago School (see Duncan and Schnore 1959 for further details).

Following this lead, Hawley (1950) divided the environment into the organic and the inorganic (for further review see Thomas 1992; Albrecht and Murdock 1990). This is akin to Park's (1936a) division of the environment into the biotic (organic) and the cultural (inorganic). In a further revision, Hawley (1986) categorizes environmental elements as according to biophysical and ecumenical classifications, which include both constant and variable conditions. For instance, the land form, flora, and fauna were considered constant biophysical elements, while variable biophysical elements include eruptions, storms, swarming of populations. Likewise, accessible human groups and exchange relationships are considered constant ecumenical elements and the visits of strangers, migration patterns, and information flows are considered variable. Ultimately, Hawley is interested in the constant elements as indicators of the existing system and the variable elements as potential causes of changes in the system. As one can see, the biophysical encompasses the actually physical environment in which the aggregated population exists, while the ecumenical class consists of the culture of those living on, and with, the biophysical environment. Thus, just as Park was very deliberate in his defining of human ecology as being distinctively sociological, and thus not general ecology, given this inclusion of the role of human culture (i.e. Hawley's inorganic/ecumenical classes), Hawley carries forward this sociological distinction with his own reversionary approach.

Hawley's human ecology paradigm consisted of three basic propositions: "(1) adaptation proceeds through the formation of interdependences among the members of a population; (2) system development continues, ceteris paribus, to the maximum size and complexity afforded by the technology for transportation and communication possessed by the population; and (3) system development is resumed with the acquisition of new information that increases the capacity for the movement of materials, and messages and continues until the enlarged capacity is fully utilized.

These may be characterized as the adaptive, growth, and evolution propositions, respectively" (1986: pp. 7). Ultimately, Hawley's propositions are aimed at understanding both the organization of human populations and the interactions of the human condition and societal evolution.

Hawley's first proposition mentions that adaptation is a direct product of the development of interdependencies among a population. As such, the proposition is laden with conceptual traces of Durkheim's "Social Morphology" and Park's theories concerning competition and dominance. Hawley's first proposition further states that "adaptation is realized with the establishment of a viable relation between population and the environment" (1986: pp. 7). Thus, the relationships between the biophysical and ecumenical classes are essential to the development of a community in any given geographic context. Put more simply, the development of any single community of aggregated individuals consists of a conscious interaction between humans and the environment in which they are staking their claim. The ability of humans to "adapt" to their environment is further explained as being a process of organization that provides the foundation for individuals to act as a single unit.

Hawley's second proposition is aimed at setting a limit to the size that any human system might grow under the static conditions that exist at the time of system development. Here Hawley specifically highlights the role of communication and transportation as contributors to the growth of the maximum carrying capacity of an environmental context. Thus, in any system there exist inhibitors of growth and checks of size that are directly related to the environmental context in which that system exists. Ultimately, Hawley understood that the ecumenical conditions that separated humans from other living organisms in their ecological systems were what would allow for further evolutionary growth. Moving from a static understanding of contributors to system growth to a dynamic one in which future advancements in technology are seen as tool through which the limiting impact of the environment might be overcome by the human condition. This is elaborated upon in his third and final proposition.

In regards to Hawley's third proposition, the adaptation of populations to a given environment is not considered an outcome, but instead an ever-evolving process. Again, this proposition is laden with ideas from Park's human ecology. Park (1936a) borrowed heavily from the ideas of Darwin in his perpetually-evolving theory of communities based on the theoretical concepts of competition, dominance, and succession (as outlined above). In fact, Park (1936a) believed that it was the application of a sociological principle, "competitive cooperation", which led to the first formulation of a theory of evolution (pp. 2). Thus, Park would believe that "Social Darwinism" provides the foundation for a "Biological Darwinism", but it is also evident that Park's interest in the evolution of aggregate populations was one at the intersection of Darwinism in both its social and biological form.

In turn, Hawley's third proposition was aimed at understanding the evolution of society, in part as a function of the development of a proper system in which biophysical and cultural conditions merged appropriately, but also in relation to the advancements that culture may allow humans that is not allowed in all systems. Thus, the human condition, most notably through culture, builds upon the interdependencies developed among members of the population (proposition 1) and the environmental limitations imposed by the specific geographic context in which the community was system existed (proposition 2). This allows for the development of advancements in technology that further allow for the human system to persist in environmental conditions that it may not have been able to previously. Here we see the distinctive nature of human civilization. Hawley (1986) believed that the system could grow to what would be maximally supported by the environment, but which could grow in an evolutionary fashion as advancements in technology allowed. The remainder of Hawley's (1986) *Human Ecology* introduces the basic assumptions of his theoretical approach to understanding the development of human systems and the assumptions that underlie that development, maintenance, and change.

The following sections each represent a corresponding chapter to Hawley's *Human Ecology* and the content focuses on the major assumptions presented in the chapter, all of which are aimed at supporting the three propositions this human ecology paradigm as presented previously in this section.

4.4.1 Environment, Population, and Ecosystem

The focus of this section of Hawley's book is on the establishment of a human system and the initial interactions between populations and the environment. First, Hawley points out that populations will fill all available space in a given ecosystem as the population organizes and that the process of human adaptation to an environmental context is the result of collective behaviors. These collective behavior processes take two forms, symbiotic and commensalistic, each of which provide access to the environment. The symbiotic sector is made up of an initial assumption that all members of a given system need access to the environment to survive. Most directly in this line of reasoning is the ability to procure sustenance. However, the collective behaviors that result in the development of a population size system are ultimately the results of social differentiation. As a product of differentiation, Hawley points out that not everyone needs direct access to the environment. In fact, some may only need access indirectly and through connections to the *n*-th order. This creates a hierarchical system of access to the environment that is related specifically to ones place in that order. For instance, Hawley points out that processors must have access to the environment through producers, thus that are not in need of direct access but instead in need of access to those that do have direct access. However, Hawley also points out that there is a degree of power associated with being the "key function", or the function with direct access to the environment to which others depend. One could easily come up with many current examples concerning the food system, natural gas extraction, and many other examples of modern hierarchical dependencies of access to the environment.

In order to put this in perspective, Hawley points out that in the most extreme of cases everyone would have the same access to the environment, leading to self-sufficiency and the non-existence of a human system. Only as a population differentiates tasks and hierarchically allows access to the environment does an ecosystem become established. Furthermore, as functions in the system become further removed from the environment there power in the system diminishes. This power is further hierarchically related to the degree of specialization associated with the function. Thus, as functions become less specialized they are likely to be carried out at a greater rate and thus decrease the power associated with the function. In this sense, the specialized function that serves as a key function with direct access to the environment is considered the most powerful. This measure of power is important given the degree of exchange in compensation associated with the differential involvement of individuals in specific functions in society. Highly powerful key functions can demand a large amount in exchange for performing the function to which the system is dependent.

Unlike the hierarchical structuring of the symbiotic sector according to access to the environment, the commensalistic sector is a horizontal categorization of individuals within a given system according to similar characteristics. According to Hawley, this categorization takes many different forms given the many different characteristics of individuals which may provide the foundation for their relationships. Ultimately, the development of these groups leads to the ability of individuals to form groups with the ability to organize for collective action. In the most extreme case, individuals of a given system are likely to understand the benefits of differentiation and social organization and thus are likely to join together in the collective action of system development and maintenance. Of course, within the system many smaller groups are likely to develop based on differential social characteristics; however, their shared characteristic of being part of the system is ultimately the most important in the establishment of the community. It is in this way that the system is able to establish a collective set of rules, laws, procedures, and sanctions. The ecosystem itself is considered formed when specific functions become responsible for the mediating the relationship between all other functions and the environment (symbiotic system) and all individuals in the system join together in unison to support the system as a single population of similar characteristics.

Finally, the continued development of an ecosystem necessitates the development of transportation and communication for environmental access. In this case, the amount of time functions have to be preformed is limited by the amount of time humans have in a day. As a result, Hawley point to the creation of specialties of time management, communication dissemination, and transportation as essential to the continued development of any population system. At this point the roles of infrastructure for the physical mobility of people and things, as well as the development of more advanced technology for the passage of information, become especially crucial. Hawley points out that another important factor in reducing time is proximity. In line with the thinking of the Park and Burgess (1925), and those to whom they drew their inspiration, Hawley believed that the development of the city was a shining example of the importance of proximity and the natural distribution of population across the city in accordance with physical distance to specific system functions. Hawley mentions that this section is primarily intended to serve as a definition of a system from which the following sections will build.

4.4.2 Ecosystem Change: The Concept

According to the propositions put forward by Hawley, one of the most central concerns of his theory of systems is the fact that they evolve over time. As such, Hawley points to the systems evolutionary development as a synthesis emerging from the influence of external forces on the already consistent present conditions. The system change model is summarized by Hawley (1986: pp. 59). Hawley points to the equilibrium stage as the steady state of the system. He notes that exogenous influences, such as new information regarding mobility/productivity, cause a series of first-order effects. The new information should be thought of as technological advancements and the first order effects of these advancements in communication and transportation are described as reduced costs (financially and socially) and an increase in the knowledge base. Ultimately, this decrease in cost and increase in knowledge result in the development of second order effects, such as the physical growth of the system (in areal terms), increase in population, and an increase in differentiation among the population due to higher levels of specialization that accompany the technological advancements. This then leads to a state of "new partial equilibrium". In this state, the population growth stops and becomes stationary, costs of transportation/communication continue to decline, knowledge exceeds its utilization, among other consequences. As the system steadies in its new, larger and more complex state, it returns to the initial state of equilibrium from which future exogenous influences will further develop the system.

4.4.3 Cumulative Change: Growth and Evolution of the Ecosystem

Once the concept of system change has been identified, it is possible to move forward with the assumptions that underlie that change. In this section, Hawley highlights such assumptions through a presentation of causes and limiting factors associated with system hierarchy, population increase, and the degree of specialization. According to the first assumption, the hierarchy of the system develops from the interaction between the system's population and the environment. Hawley points to the existence of complex units in this portion of his theory. He makes the point that every specific function can be coupled with other system functions, creating a higher order function, which is carried out by a complex unit. These complex units can be either *corporate* or *categorical* and they are further divided by a unifying principle. In the sense that the complex unit is general to the entire population of a given system it is a corporate unit.

Hawley points out that the best example of a complex unit would be the household in that it is a miniature system inside of which a further division of labor exists in order to efficiently procure sustenance, train the young, and provide many other services. A categorical complex unit on the other hand is one that is specific to a group with a shared set of characteristics. Hawley presents the example of a neighborhood in which only members associated with that neighborhood feel the attachment to it. This differs from the corporate unit because everyone in the system experiences their own generic version of a household (corporate unit) but only those with a shared interest in a specific area of the city experience a specific neighborhood (categoric unit). Hawley further differentiates between the corporate and the categorical given the unifying principles of familial, territorial, and associational characteristics (1986: pp. 74). For instance, examples of corporate units in the familial would be the household, while an example of categoric units in the familial line are kin, clan, and tribe. Similarly, corporate units in the territorial line include villages and cities, while categoric units in the territorial line include neighborhood, ghetto, and ethnic enclave. Finally, example corporate units in the associational line include industry, retail, school, and government, while categoric units include caste, class, club, union, professional organization, etc. The functional specialization that occurs as higher-ordered functions among these units helps to contribute to a hierarchically structured system based on access to the environment, and related decreases in power due to increases differentiation and specialization. Furthermore, Hawley points out that the systematic control of capital and resources leads to higher levels of relative depravation at each lower level in the hierarchy. At this point the categoric unit often exhibits its power through the collective action based on its opposition to some systematic inequality to which the unit is exposed.

In regards to the effects of cumulative change on the population, Hawley points out that cumulative change of a system entails increased size and complexity. The previous section highlighted the change model, which points to the introduction of technological advancements as a driving force of a larger system in terms of geographic space and population. As such, the cumulative change of a system introduces increasingly higher levels of differentiation and calls for a larger population to staff newly created positions aimed at managing the functions of the increasingly complex and large system. Given that relationship, population size grows to a point at which all positions are filled and allows for higher levels of support for specialized tasks.

4.4.4 Cumulative Change: Expansion in Time and Space

Following the introduction of consequences associated with the size and specialization of system populations as a result of cumulative change, Hawley introduces propositions associated with the growth of systems. In particular, he is interested in variations in that growth that are related to mobility "costs." At this point Hawley reviews the traditional patterns of growth that systems have undergone as they increased in size and complexity. The center expansion pattern that resulted in as a product of advancements in transportation is one that is familiar to many sociologists, and by this point in this book, it is becoming a frequent point of reference. The most famous depiction of the center expansion is the concentric zonal model of Park and Burgess (1925). One may also think back to the economic geography model of

Von Thunen (1826). In this type of spatial organization, the center of the system (city/community) is the focal point of all of the systems most primary functions. As one move away from the center of the system, land depreciates, areas become increasingly residential, and open spaces become more abundant.

With the advent of railways and personal transportation, such as automobiles, the system continued to expand outward, radiating from the center. Hawley points out that this is the result of holding all mobility costs constant and tends to be representative of the preindustrial city. As transportation advancements become more affordable, the clustering of functions in the central city disperses into homogeneous functional enclaves. Ultimately, Hawley moves into a discussion of the deconcentration pattern of system growth that develops in regards to the ease of access to transportation and communication from greater distances than were previously allowed. This leads to a series of events that underlie the transition of previously independent systems into a larger system made up of geographically proximate systems.

One might think specifically of the Boston-Washington megalopolis, stretching from Washington, DC in the south to Boston in the north. At a point in time, all of the cities that are currently considered part of this "super-city" were independent cities serving a population specific to that system. Now there are cases individuals residing in one city and commuting to work in another that is over 100 miles away, as is the case in New York City where train loads of workers pour into the city everyday from states that do not even border New York state in some instances. In another case, one could think of the U.S. itself as a merged set of previously independent systems; and on the extreme one might even conjure ideas of world's systems theory, considering the global environment of today's world as being a merged set of previously independent systems. Hawley saw this as the merging of systems in the creation of a single larger system. Specifically, he pointed out that as populations deconcentrate and systems grow larger, previously independent biophysical environments would become one larger biophysical environment. As a direct consequence, distinct cultural groups are also swallowed by the growing system in favor of a single more diverse ecumenical system.

4.4.5 Limits to Cumulative Change

Finally, in the last section of *Human Ecology*, Hawley deals with the limits of change. Hawley is very specific in the presentation of a single assumption; only the efficiency of transportation and communication may limit the complexity of the system. Regarding this efficiency, Hawley highlights the well-documented growth in communication and transportation advancements as being related to the exponential increases in the scale and complexity of the human ecosystem. Hawley points to the relatively meaningless borders of international states and the continued development of a world-systems theory. In this sense, the growth of the human system that was documented in the previous section is thought to have reached a point, or near reaching this point in 1986, where the independent systems of the world would not longer operate independently; but instead part of a larger global system of interrelated parts. Finally, the question of what might limit the advancement of scale and complexity in the human ecosystem; to which, once again, Hawley believes is linked only to the efficiency of communication and transportation.

4.5 Chapter Conclusion

As Hawley pointed out, the ability to continue to advance levels and mediums of communication and transportation have continually increased the scale and complexity of the human ecosystem. Ultimately, it is this human condition that separates the study of human ecology from the study of a more general ecology. This is the same distinction that Small and Vincent (1894), Hayes (1908), Park (1936a, b), and many other human ecologists since then have made. Still others believe that the study of human ecology is more a sub-field of ecology, geography, or other academic disciplines. We believe that human ecology is a part of a geographical sociology, or a geo-sociology, much the same way as social psychology has given us a psycho-social approach to understanding the interrelations of the mind, self, and society. Instead, a geo-sociological approach is much more interested in the primacy of neighborhood and environmental effects but does not preclude the Durkheimian idea of social reconstruction through micro-level change.

The work of the early human ecologists that was presented in this chapter laid much of the foundation for all future theoretical developments that have been made and continue to be made in the relationship between humans and their ecological context. However, the sub-field of human ecology itself no longer holds the same place it once did in the field of sociology. Perhaps part of that is due to the fact that there are so many different meanings for the term, across so many different disciplines, that it no longer has the same distinctly sociological flavor that many of its founders had intentioned. Human ecology ultimately fell out of vogue with sociologists in the late 1950s early 1960s (Gross 2004). Leo Schnore (1958) identifies a declining trend in the field of sociology concerning interest, and self identification, with human ecology among members of the American Sociological Association. However, the late 1960s saw an increase in human ecological analyses under a concerted effort to respond to growing environmental problems (Gross 2004). This new ecological movement was couched within the development of environmental sociology. We argue that the new spatial methodologies facilitate the enhanced use of human ecology as a theoretical framework. The next chapter of this book examines the development of and focus on environmental sociology as part of a larger survey of geo-sociological developments over the last quarter of the twentieth century.

Chapter 5
Contemporary Movements in Theories of Location

5.1 Introduction

Recently, increased attention has been given the effects of social context and ecology concerning many of the core interests in sociological research. Spearheading this recent interest is the technological advancements in computational power and sophistication that are now widely available to most researchers in most educational institutions in the country. Chapter 7 will introduce many of the existing quantitative methods that have been developed and become popular as a way of putting social phenomena and human behaviors into "place".

At this point it is important to note that theoretical advancements are developing but at a much slower pace. This is to be expected in some sense. As Hawley's theory of system development itself posited, at the point of rapid change, knowledge accumulation occurs at a much faster rate than does its utilization. Today we see the extremely rapid advancements that have taken place in regards to statistical programming and quantitative methodologies, which in many cases are unable to be effectively implemented due to a lack of a theoretical understanding of the processes that some methods are designed to study. For instance, the ability to track diffusion and physical mobility of populations and ideas is available using various statistics of spatial clustering (see Cohen and Tita 1999; Porter 2010 for examples). These methods are certainly in need of refinement, but this can only occur through their usage alongside of theoretical development.

This chapter introduces some of the contemporary theoretical advancements that have been made since the foundations of human ecology and the early interpretations of geographical sociology were presented (primarily in regards to the work of Hayes 1908; Park 1936a, b; Hawley 1950). These advancements include the establishment of an environmental sociology paradigm by Catton and Dunlap, the 'New' Chicago School led by Robert Sampson and his colleagues; and the views of the oppositional Los Angeles School led by Micheal Dear's postmodernist theories of location. There are of course other developments in the fields of community studies, urban/rural sociology, demography, criminology, and epidemiology that continue to

J.R. Porter and F.M. Howell, *Geographical Sociology: Theoretical Foundations and Methodological Applications in the Sociology of Location*, GeoJournal Library 105,
DOI 10.1007/978-94-007-3849-2_5,

take shape within the field of a geo-sociology. Given the scope of the current chapter, we only focus on the before mentioned topics for a review (see also Howell and Porter 2010).

Furthermore, we conclude the chapter with a focus on what needs to be theoretically developed for a geo-sociology. In particular, what is *not* theoretically understood in regards to the relationship of social processes, human behavior, and social organization to specific spatial concepts is key. For instance, what does it mean theoretically that we can place a series of individuals within varying contexts as we do in the implementation of hierarchical linear and mixed modeling regression techniques? At this point, there is an explicit interest in the *containment* of some focal outcome within a specific ecological context, which likely varies for other members in the same sample. What does this suggest theoretically? Furthermore, why is it important to test for the effect of *adjacency* and *proximity* when sociologists implement models of spatial regression with neighborhood weights and through the employment of statistical techniques aimed at identifying "hotspots" and "significant spatial associations"? What constitutes a spatial neighbor in the first place? And what *order* of contiguity must be taken into account? All of these are questions that remain mostly unanswered today due partly to the recency of spatial methods being adopted in sociology.

Ultimately, the indicators of containment, intersection, proximity, adjacency, and other spatial concepts are likely to provide the general core of what a geo-sociological approach should be interested in understanding. These spatial relationships provide the underlying foundation from which middle range theories, such as Sampson and colleagues (1997) collective efficacy, can develop (see also Howell 2004). An understanding of how and why collective efficacy impacts individual probabilities to commit violent crime has been well advanced by the researchers, but a more general modern understanding of why the containment of an individual within a unit that displays a certain characteristic at the aggregate level continues to be less advanced. With the continually increasingly interest of sociologists, criminologists, demographers, and many other disciplines in environmental and neighborhood effects, perhaps we are still in the process of collecting enough information to do so.

5.2 Environmental Sociology

Gross (2004) points out that throughout the 1960s, the interest of sociologists moved away from the ecological explanations but were invited to return with the increasing attention being given by the public to environmental concerns. In response, the 1970s saw a refocusing of issues of the environment by sociologists. In particular, the establishment of an environmental sociology section of the larger American Sociological Association and the influential articles on the topic of environmental sociology that were further introduced by Riley Dunlap and William Catton (Gross 2004).

The introduction of the *New Environmental Paradigm* (NEP), by Catton and Dunlap (1978) was driven by a desire to overcome the limiting factors of existing sociological theory; a theory littered with explanations of social processes driven by social facts. According the Catton and Dunlap (1978), it was the focus of sociological theory on social facts that had limited researcher's ability to explain the relationship between events not social processes due to the narrow scope of what was to be considered a "fact". This anthropocentric view of human behavior, social organization, and other social processes did not allow sociology to deal with the non-anthropocentric "facts" that impact social life (Catton and Dunlap 1978). Catton and Dunlap further point to the relative neglect of human ecology to take into account the physical environment as an influence on social organization (cited from Michelson 1976). They instead highlight the human ecological approach as being closely tied to the spatial and the social as indicators of their relegation to the *Human Exceptionalism Paradigm* (HEP), from which the NEP paradigm had hoped to overcome. The new paradigm eventually grew to be very inclusive of a large number researchers from many sub-disciplines in the field of sociology (for a review see Dunlap and Catton 1979a, b).

At this point, it is important to review the basic assumptions of each approach, which are presented by Catton and Dunlap (1978) in order to better understand the theoretical tenets advanced by the NEP. The basic assumption of the HEP state that "(1) Humans are unique among the earth's creatures, for they have culture; (2) Culture can vary almost infinitely and can change much more rapidly than biological traits; (3) Thus, many human differences are socially induced rather than inborn, they can be socially altered, and inconvenient differences can be eliminated; and (4) Therefore, cultural accumulation means that progress can continue without limit, making all social problems ultimately soluble" (pp. 42–43). It is easy to see the focus of the HEP paradigm resting on the core factor of culture, which is considered by followers of the HEP approach to be the deterministic factor in human relations and social organization. In contrast, the basic assumptions of the NEP state that "(1) Human beings are but one species among the many that are interdependently involved in the biotic communities that shape our social world, (2) Intricate linkages of cause and effect and feedback in the web of nature produce man unintended consequences from purposive human action, and (3) The world is finite, so there are potent physical and biological limits constraining economic growth, social progress, and other societal phenomena" (p. 45). Here Catton and Dunlap highlight the theoretical movement away from human exceptionalism and towards humans as a single group in the much larger ecosystem consisting of finite limits.

The authors further support the introduction of the NEP approach through the presentation of undeniable environmental facts, which they assert are essential to the understanding of social facts. To illustrate, Catton and Dunlap present a series of examples associated with the utility of the NEP approach in regards to temporally-relevant issues of social stratification. Such a linkage of natural resources to the social stratification of humans has a long history. Perhaps the best summation of the relationship was presented by Franklin Thomas (1925) who asserted that "a rich or poor environment affects man indirectly by .... determining its composition and general character" (p. 120). As an account of the historical relationship between humans and the environment, Thomas references the ancient ideas as presented by Hippocrates, Herodotus, Thucydides, and others in reference to the relationship between population characteristics of ancient populations and differential bounties of sustenance, natural resources, and land. Other attempts to understand the relationship between social organization and the finite environment had been made about the same time as the development of Hawley's *Human Ecology*, although it is interesting to note that most were by geographers (see Anderson 1951), whom have a distinct interest in the relationship of the physical environment to culture on the social or human geography side of the discipline.

It is at this point that one must take the time to think about the disciplinary overlap of sociology and geography when the ideas of environment are brought into the discussion. In regards to the simple relationship between humans and the soil, as are presented by Thomas (1925) and Anderson (1951), we should consider them to be human geography. However, at the point of which the relationship becomes more complex and it is the social relationships that are of the primary focus, we should consider these issues of geo-sociology. The focus of Catton and Dunlap's research directly takes social relationships as their primary focus, only including environmental factors as a variable that help us to understand the social organization and processes that lead to social inequality. However, the somewhat dismissive roles of the basic assumptions of the HEP approach are also to be understood as incomplete in the development of a paradigmatic approach. In asserting the assumptions of the NEP approach listed above, the environmental sociology approach removes too much of the HEP approach. Specifically, the basic tenets of the NEP approach do not include a link to human culture. While a dominant focus on culture may narrow the focus of sociologists in their ability to explain and understand modern day relationships of the social organization of humans within the larger ecosystem (Catton 1994), it should most certainly be taken into account as an exceptional component of the human condition.

In this respect, geo-sociology takes the assumptions of both. A geo-sociological approach makes the following synthetic assumptions: (1) environmental facts help to shape the characteristics and organization of human populations and these facts are based within a finite world of limitations to growth, expansion, etc., (2) while humans exist as a single species in a larger ecosystem, humans do have inherent in their organization and development an exceptionalism characteristic of culture which allows for the application of technology to extend beyond what other species may be able to with similar resources. Thus, a geo-sociological approach takes into account both finite limitations, environmental factors, and the human exceptionalist characteristic of culture. This is not necessarily a new idea. Within Hawley's *Human Ecology* the ideas of adaptation and change can easily be linked to the environmental context in which functional order in access to the environment, along with specialization, is the distinguishing characteristic of power, and thus social stratification.

What the environmental sociologists add to the discussion is the fact that the functions which access the environment are likely to dry up in the sense that they are finite. As such, the foundations for power and stratification will shift, or even become exaggerated, at the increasingly limited access of groups to the natural resources which the earth provides. Thus, it is neither the restrictions of the earth's resources, or the inherent capabilities of culture to overcome societal hardships through technological advancements, but instead a combination of the two in which human beings are likely to adapt at an accelerated pace compared to other species in the ecosystem because of the foundation of culture. Again, this occurs within a finite world and in an ideal state would necessitate the adaptation of power dependencies, food sources, etc. it order to meet the needs of populations that are likely to use all of the available resources in the current system of providing fuel, sustenance, and raw materials. As Catton and Dunlap (1978) point out, the movement towards a stead state in the ecosystem is the common goal of all of these environmental/ecological approaches.

In a very Malthusian vein, perhaps the self-checks of decreasing fertility rates, and the associated lag of tailed population growth later this century, are an adaptive approach to reconciling such a relationship. Further, technological advancements in the way of non-fossil fuel transportation would also provide a freedom from the restriction of natural gases and oils. Perhaps, it is these types of adaptations that will continue to occur until a steady state is reached. Although, the environmental sociologist is likely to remind us that once a finite material is gone, it is gone for good. In these cases, the relationship of man to his environment is only to become more complexly linked as the human species continues to adapt to the ever-evolving role they play in the larger ecosystem.

We agree with the argument of Buttel (1986) in his assessment of the place of environmental sociology within the discipline of sociology itself. That is, environmental sociology has secured itself as a niche within the larger discipline, but does not do justice to ecological analysis as a replacement to other ecologically-centered theories. However, the attention drawn to the natural environment does compliment quite well the existing theories of social organization. As such, environmental sociology stands unified as a core area of sociology; a geo-sociology. Geo-sociology, then, is to be considered the macro in the sense that it concerns itself with population scale organization and system level phenomena; however, it should also be considered meso in the sense that it concerns itself with the spatial relationship of individuals to structural conditions. Finally, it moves beyond the simple tag of macro- or meso-sociology is that its primary concern is the spatial relationships that exist among both levels of analysis in accordance with the concepts of containment, adjacency, intersection, proximity, etc. It is important to note here that this includes spatial relationships of both physical and social characteristics as being determinant of social organization. As such, a number of middle range theories have developed over the last couple of decades in regards to relationships between humans and their environment. The next section will introduce one such relationship that finds its connections to the Chicago School, but makes a self-exclaimed stake the "New Chicago School".

5.3 Middle-Range Theories of Human and Ecological Relationships

Within the last decade, Robert Sampson and colleagues have strongly influenced an increasing interest in the ecological context in which variations in the human condition manifest themselves (Sampson 2002, 2003, 2008, 2009; Graif and Sampson 2009; Sampson and Sharkey 2008; Morenoff et al. 2001). Sampson (2002) argues that this can ultimately be seen as a return to the foundations of sociology, which were originally forged in the widespread social changes taking place at aggregate levels in the times of the industrial revolution. The article goes on to highlight the historical influence of the Chicago School of sociology, both in regards to its longstanding focus on context/place as a determinant of individual social behaviors and also, but perhaps more importantly, as a stand-alone unit of analysis for understanding the changing nature of populations in aggregate form as a result of variations in contextual conditions (Sampson 2002; see also Coleman 1994; Abbott 1999). Thus, it is only appropriate that one of the major contributions Sampson (along with Raudenbush 1999) has had in pushing forward the Chicago-style approach to social research is the introduction of an “Ecometric” methodological concept to scientifically assess ecological settings and their impact on various outcomes in both individual and aggregate form.

Most notably, the research of Sampson and colleagues has focused on neighborhood studies, primarily within the Chicago area. However the application of the ecometric approach has been implemented across a wide-range of geographic settings. In the work referenced above, much of the data builds on primary data collected as part of the Project on Human Development in Chicago Neighborhoods (PHDCN). While some of the work related to this study was concerned with the variable conditions of social context in regards to observable variations in human capital (collective efficacy) and its impact on the occurrence of violent crime, the explicit focus of the work itself was on neighborhood’s as conditions in which more general social behaviors occurred (Sampson et al. 1997, 1999, 2005). As a result, the measurement of aggregated human capital, and its associated relationship to the positive implications of high levels of collective efficacy, took place at a neighborhood level. The neighborhood indicator of efficacy could then be examined as a dynamic and fluid indicator of the ability to maintain order through shared expectations for action, in direct relation to the accumulation of human capital at an aggregate level (i.e. social capital) (Sampson et al. 1997; Sampson 2003).

Here we get a glimpse at what it means to be *contained* within a given ecological context. The spatial relationship is of individual survey respondents that have a direct relationship to a neighborhood. In this sense, the neighborhood should be thought of as a polygon and the individual as a point contained within the polygon (further explanations of these images are presented in the following two chapters). From Sampson’s work, we see that distinct differences exist between polygons in regards to levels of collective efficacy, and that these distinct between polygon differences directly influence the behaviors of the points within them. In the real world, we understand that neighborhoods are different and each provides a specific ecological context in which individuals develop. Each of these specific ecological contexts are further related to cultural orientations or what Hawley called "inorganic/ecumenical" components of the environment that directly influence the local sub-system. In the case of Sampson's work, the sub-system of the neighborhood contained a series of specific characteristics, one being "human development," that were found to be directly linked to individuals propensities to commit criminal behaviors.

This re-discovery of the "neighborhood effect" has become increasingly prominent within the social sciences (Sampson et al. 1999, 2002; Raudenbush and Sampson 1999; Sampson and Sharkey 2008). Specific to the examination of criminal offending, a new approach to understanding the social processes that link disadvantage to crime emerged. No longer does it suffice to say that crime occurs at higher rates because communities are disorganized or because of the fact that the opportunity structure allowed for higher rates of crime, as posited by the social disorganization and routine activities theoretical frameworks respectively. Instead, a new paradigm has emerged which makes a focused attempt to understand the underlying community processes that exist as a product of such conditions (in varying degrees) and in an endogenous/exogenous fashion also lead to individual variations in rates of criminal offending (Sampson 2002). This focus on the aggregated development of capital, and associated levels of collective efficacy, has coincided with a much larger movement back to ecological examinations of social processes and group behaviors. The ideas harken to a time of C. W. Mills and *The Sociological Imagination* (1959), when the placement of social action within the spatio-temporal context in which it occurs was central to understanding the mechanisms driving the action.

Here the focus is on location, in particular, the geo-sociological imagination (Porter 2011), and even more so, on the spatial relationships that link place and individuals to their larger social and physical environment. As this section has highlighted, the containment of an individual within a neighborhood is a single example of a spatial relationship. Many others will be further introduced in the following chapters. The next section of this chapter moves forward in an attempt to explore an even more complicated relationship between social organization and space as presented by what has been called the "Los Angeles School". Most prominent to this movement is the work of Micheal Dear, who presents a postmodernist theory of location in regards to contemporary social organization.

5.4 The Los Angeles School of Urbanism

As with a most theories and ideas throughout academia, spatial theory and related modern theories of location have become threatened by the recent rise in attention given to the postmodern movement. One of the leading writers on the postmodern movement, within theories of location, is geographer Michael Dear from the comparatively lesser known Los Angeles School (in relation to the Chicago school).

Although Dear is a geographer, his primary interest is that of social geography, or human geography, and the relationship of space to social theory. The development of his writing coincided with the larger trend of giving increasing importance to geography along with many of the political, economic, and environmental issues that arose during the time period (National Research Council 1997).

Dear's (1988) approach within the postmodern movement is to reconstruct theories of location. He goes on to explain the effect of such a process as: (1) repositioning geography to have a central position within the social sciences, (2) recasting the internal structure of the discipline (geography), (3) re-forging geography's links with the mainstream debates in the philosophy and method of the human sciences. In integrating human geography and social theory, he is concerned with the role of social theory as the "illumination of the concrete process of the everyday life" (Dear 1988). Dear (1988) was interested in the development of geographic areas as they relate to postmodern thinking. Dear believes that, as modern cities (i.e. Chicago) developed around concentric zones tied closely to primary transportation hubs, newer postmodern cities developed in a much more random pattern tied only to non-physical communication hubs (i.e. telecommunications, etc.). Ultimately, this development calls for a movement beyond basic ecological and environmental explanations for society in simple terms of spatial proximity and towards an analysis of social geographies.

Dear goes on to note that the use of history in the social sciences as a mechanism of time for examining behavior is much farther along than the use of geography as a way of explaining behavior based on space or location. He further makes the claim that by fully understanding the potential to which the use of human geography can help develop social theory, geography itself will re-situate itself in the center of a newly defined paradigm on human inquiry. Dear explains that human landscapes are created by knowledgeable actors (agents) operating within a specific social context (structure), and that the structure is transformed by the agents making any narrative on the human landscape an account of the reciprocal relationship between long-term structural arrangements and short-term practices of individual agents.

This last statement shows the degree to which human geography is linked to social theory. Further proof of such a relationship is that social relations are constituted through space, constrained by space (boundaries), and mediated through space (Dear 1988). The use of location in social theory is an exercise in reflexivity in which any single locale is a complex synthesis involving the ever-evolving social processes and their relation to the above-mentioned, location-specific limitations (Dear 1988).

Dear's argument takes aim at urbanism from a postmodern point of view. In doing so, he compares the Chicago School's Concentric Zonal Hypothesis with that of a newly developing postmodern view of the city of Los Angeles from the aptly named Los Angeles School. If one recalls, the Chicago School view of the city was one of a central business district and concentric zonal rings, which dispersed outward forming layers of rings (see Fig. 2.1 in Chap. 2). Each of these rings, then, constituted a different neighborhood that could be characterized by the type of housing, crime rates, social class, etc., which existed within the ring.

The Los Angeles School model uses a postmodern method of deconstruction to show that the Chicago School model is outdated and of little use anymore. The Los Angeles School model is not perfectly situated in concentric zones; it is instead a random layout of fundamental urban characteristics that more aptly make up the postmodern urban center. The Chicago School assumptions of uniform land surface, universal access to single central city, free competition for space, and the notion of outward development can be dismissed from a postmodern point of view, as they simply do not represent current day reality. The development of the Los Angeles model, then, can be seen as the evolution of spatial theory taking place as a result of deconstructing theoretical assumptions and reexamining the sources of current knowledge.

Ultimately, the thinking of the Los Angeles School provides a framework from which advancements in communication and transportation can be understood in their re-shaping of the city. Furthermore, the ideas presented by Dear provide a framework from which spatial contexts can be compared. For instance, why is it that we suspect a contiguous nature of diffusion versus a hierarchical nature. In contiguous approach, one would expect that population and ideas disperse from a polygon (per the earlier illustration) to a neighboring polygon. However, others who study diffusion would argue that many processes spread hierarchically and in relation to ecological characteristics. Thus, in developing a theory of spatio-temporal mobility, one should consider not only the spatial relationship of adjacency/proximity, but also the inorganic or ecumenical characteristics of both the origin and potential destination locations (for a recent review of diffusion processes see Porter 2010).

5.5 Recent Developments in the Integration of Spatial Thinking in Sociology

During the 1990s, a number of social scientists—led by geographers—argued for the spatial integration of spatial thinking throughout the social sciences (Goodchild et al. 2000). Candidly, we believe that as a discipline sociology was far behind the widespread adoption of these methods into the curriculum of graduate training. In 1999, the National Science Foundation funded the launch of the Center for Spatially-Integrated Social Science at the University of California-Santa Barbara (http://www.csiss.org) under the guidance of Michael Goodchild, a leading geographer specializing on spatial analysis and Richard Applebaum, a sociologist. Later, the "teach spatial" website at UCSB (http://teachspatial.org) focuses on this sorely needed training effort. This effort has provided a tremendous boost toward the integration of spatial thinking across the social sciences, including sociology.

A collaborator, Luc Anselin, now at Arizona State University (http://geodacenter.asu.edu), had been plowing new ground since his dissertation work at Cornell was published as a breakthrough text (Anselin 1988). Anselin's cadre of collaborators have been developing freely-available software through which spatial analysis techniques—especially the challenging spatial regression—can be conducted on personal computers. Groups of social scientists in the United Kingdom (Ripley, Diggle, Fotheringham, Bailey, Besag, Haining, Openshaw, Lawson) and elsewhere in Europe (Bivand) were also pushing the analytical techniques forward to handle spatial data in the social sciences. The GeoVista Center at Pennsylvania State University also made significant inroads toward the development of new ideas of visualizing spatial data, especially incorporating the temporal dimension.

Actual training in sociology was sparse at the time as only three graduate courses were known by these authors to be taught in 1997: Frank Howell at Mississippi State University; Paul Voss at the University of Wisconsin-Madison; and Joe Francis at Cornell University.

As Howell (2004) pointed out at a session on spatial analysis in rural sociology at the Rural Sociological Society annual meetings, the gap between spatial theory and methods was in need of a "few good theories of the middle-range". Rather than throw out extant theory, he argued, it is perhaps more fruitful to modify current theory in spatial terms. As Lieberson also argued:

> ...in social science we talk about theories being either right or wrong...It is a matter not of simply rejecting a theory, but rather of evaluating a theory, knowing at some point that it will have to be modified or even superseded...we are more likely to want to destroy a theory or...worship a theory and thereby resist its change or modification. (Lieberson and Lynn 2002)

An example of such simple transition would be to modify traditional concept of "rural" as a size-of-place continuum into one involving the contiguity of places. The spatial contiguity of differing sizes-of-place then becomes part of the analytical approach. A more aggressive example is Greve's ecological theory of local density dependence in organizations (banks, in his case) as alternative to models implying spatial contagion and competition. The major point is that the taking of theoretical and methodological "chances" is predicated upon creative spatial thinking as applied to important sociological theories. In short, space is not the "final frontier" but it may have been the first one that sociology faced!

5.6 Chapter Conclusion

In this chapter we have introduced the topic of environmental sociology with a specific focus on the contributions that it early theoretical formulation had in the development of a larger theory of ecologically integrated sociology. The seminal work by Catton and Dunlap (1978) brought to light a systematic oversight of much of the existing sociological theory, the lack of attention paid to environmental facts. As the authors would point out, it was sociology's over concern with Durkheim's notion of social facts that caused scholars to be unable to attend to some of the social phenomena that were occurring as a result of their human exceptionalist viewpoints, the belief that technology could overcome all limitations on progress, and that culture was the key to man's domination of the his environment. The environmental concerns of the 1970s proved that this was not actually the case and the anthropocentric theory of sociologist was not amenable to such a deterministic nature of the environment. Ultimately, others have shown that environmental sociology has a niche in the field of sociology, but not as a replacement, or even a competitor, of macro level theories of structure. Instead, it is a compliment which helped to push forward ecological thinking with an explicit focus on the environmental facts and social facts.

Ultimately, it is our assertion here that this combination is important in understanding the role of ecology in the sociology. Together the two focuses allow researchers to understand that culture is unique to humans and does allow humans the opportunity to adapt in ways that other species cannot. However, it also brings to light the point that we do live in a finite world with fuzzy limitations of progress. Culture as a core of the human condition may allow for the progress to a certain point, but ultimately it will be the process of adaptation that sets human beings apart from other species. In this sense, technological advancement may lead the way in the adaptation of a human system as it evolves over time. This would make environmental sociology much more of a compliment to the theories of human ecology than the antithesis that it was originally developed as. Regardless, this is beyond the scope of the current book.

We are interested in presenting the theoretical foundations of a geo-sociology, with a focus on the ecological movements that have worked to establish a theory of location in sociology. To this point, we have presented many of the historically important sources from which this foundation is built. We understand that others have not made it into our discussion, but believe that we have reviewed the most influential. To this foundation, we add the importance of spatial concepts as a direction for future development of a geo-sociological theory. Early in this chapter we introduced some of these concepts (i.e. containment, intersection, proximity, and adjacency). Adding to the introduction, we believe that geo-sociological theory should be one of location and of locations in reference to one another. At this point we are moving slightly away from the early theories of human ecology (Park, Burgess, Hawley, and others) and towards a theory of locational relationships. We have provided an example of the development of one such middle-range theory that aims to understand why individuals that are contained within a neighborhood, with its own specific characteristics, are likely to be affected by that neighborhood in a way that is different than individuals from a completely different neighborhood, with its own set of characteristics. Similarly, why are sub-systems (i.e. counties, neighborhoods, cities) that are proximally close to one another more likely to share similar levels of response to outcome variables?

The same questions can be asked of all spatial relationships. To this point, we do not provide much of a response. What we hope that we have done is presented enough information to get other scholars thinking about these questions beyond the specific relationships that result as a product of our research articles. The following section of this book (Chaps. 6, 7, 8) discusses geo-social research in application through an overview of methods to spatialize and analyze social data.

Part II
Spatial Context in Social Research: Methodological Applications

Chapter 6
Making Data Spatial

6.1 Introduction

This chapter illustrates how geographic information systems (GIS) can complement micro-level survey research data.[1] We provide several examples of uses of spatial technology in surveys, introduce some GIS terminology needed by survey researchers, and discuss issues in linking GIS and survey data. Example applications illustrate the use of spatial analysis procedures with geo-referenced survey data. We identify some key barriers to incorporating spatial methods into the design, execution, and analysis of micro-level surveys and suggest approaches to dealing with them effectively.

Combining GIS with traditional survey research methods can deepen understanding of phenomena represented by survey data by revealing underlying relationships involving geographic location and spatial proximity. Such relationships may be uncovered using both visual displays and spatially-centered statistical methods, yielding statistical estimates with increased reliability and better understanding of the phenomena of interest. This chapter concentrates on GIS as a visualization tool for depicting spatial relationships, but includes a short section on recent advances in spatial statistics and their applications.

Continued advances in desktop computing have fostered a rapid increase in the availability and usage of GIS. High-quality geographically referenced images facilitate spatial displays of social data, leading to growing use of GIS across the social sciences (e.g. Goodchild and Janelle 2004a, b). GIS is best used in conjunction with research problems in which spatial elements are inherent. Such spatial

[1] An earlier version of this chapter is printed in the *Handbook of Survey Research* 2nd Edition (Howell and Porter 2010). The authors wish to acknowledge the editorial commentary of Peter Marsden and James Wright as a contribution both directly to the original version and indirectly to the current versions of this chapter. However, all errors of fact or interpretation should be attributed to the authors.

J.R. Porter and F.M. Howell, *Geographical Sociology: Theoretical Foundations and Methodological Applications in the Sociology of Location*, GeoJournal Library 105,
DOI 10.1007/978-94-007-3849-2_6,

components are often present in survey data on individuals, places, or entities that are spatially identifiable, unique, and proximally relatable to one another. Theoretical considerations will dictate which spatial elements are appropriate foci for particular applications of GIS to social survey data.

Using GIS in conjunction with micro-level survey data has become more common recently. Variation in its acceptance and development within academia is substantial. Examples in the following section illustrate some of its uses within social science disciplines, but are by no means exhaustive (for a more comprehensive treatment, see Goodchild and Janelle 2004a, b). We focus on fields that have been active in linking social phenomena to space, including epidemiology, sociology, criminology, and demography.

6.2 Previous Work Linking Survey Data to GIS

Epidemiologists have conducted much pioneering work in linking GIS and micro-level survey data. Spatial epidemiology has been a recognized specialty for quite some time. Elliott et al. (1996), for instance, provide a cogent overview of geographical methods for "small-area studies," many of which are based on catchment surveys of health issues, building on Dr. John Snow's pioneering work on the cholera outbreaks in London during the 1850s. Snow's research was one of the first to incorporate geographic images in the spatial analysis of social phenomena collected from individual-level survey instruments.[2] More recent applications of GIS in epidemiology include using individual and aggregate-level survey data to predict seasonal disease outbreaks and facilitate strategic healthcare responses, and monitoring air and soil quality as factors associated with disease outbreaks (Foody 2006). The ability to geo-reference individuals and their spatial proximity to disease outbreaks, health care facilities, and local community epidemics greatly improve our understanding of these phenomena and implementation of appropriate policy initiatives to mediate their impact.

Other disciplines have examined spatially-centered survey data to better understand other social phenomena. Sociologists have used GIS with survey data to enrich understandings of social relationships. For example, Entwisle et al. (1998) collected survey data on individuals and households in Nang Rong, Thailand. Their research team was interested in a number of interconnected individual and tribal relationships, but individuals could not be readily linked to larger ecological units because the region lacked politically defined boundaries. GIS analytical tools allowed the researchers to spatially reference individual-, household-, and tribe-level characteristics to the geographic landscape using distance-based definitions

[2] See http://www.ph.ucla.edu/epi/snow.html for details.

and Thiessen polygons (see Entwisle et al. 1998 for graphical illustrations).[3] Here, coupling GIS methods with traditional household surveys allowed examination of a larger spatial network of relationships at both the household and tribal levels.

GIS can also add value to a number of publicly available survey datasets frequently used by sociologists. Mouw and Entwisle (2006) used the National Longitudinal Study of Adolescent Health (also known as AddHealth) to examine the effects of racial and residential segregation on in-school friendships. Using a social network approach together with geo-referenced survey data, the researchers showed how out-of-school spatial proximity directly affects the in-school social relationships and networks that adolescents develop.[4] Incorporating GIS into the analysis was facilitated by the AddHealth restricted-use database including spatial identifiers. A growing number of major surveys make such information available.

Work in criminology and demography offers further examples of the utility of GIS in analyzing social survey data (see Scott and Mears 2005; Peters and MacDonald 2005; Steinberg and Steinberg 2005 for overviews). Many studies in these fields center on national-level censuses of reported criminal offending data (i.e. the Uniform Crime Reports from the FBI) and population characteristics (i.e. the decennial Census), respectively. Spatially centered methodologies link these large, publicly available, survey data to space in micro and aggregate form, helping to embed spatial thinking within emergent theoretical perspectives in those fields.

Criminologists have long sought to understand the ecological and contextual environment in which crime occurs (Paulsen and Robinson 2004). Many ecological theories of crime explicitly involve inherent spatial dynamics in the mobility of crime through processes of displacement and diffusion. By applying spatial methods to survey data, such as the police agency-level reports of offenses recorded in the Uniform Crime Reports (UCR),[5] researchers have identified patterns of geographic mobility in criminal offending at local (see Cohen and Tita 1999) and national levels (see Porter 2008). Spatially referencing aggregate levels of criminal offending to geographic and proximate locations revealed spatio-temporal patterns in the mobility of crime. Such work yields new information for local policing agencies which may be used to allocate resources to the ever-mobile phenomena of criminal offending, e.g. by identifying "hot spots" of high-crime activity.

The title of Entwisle's 2007 presidential address to the Population Association of America, *Putting People into Place* (Entwisle 2007), highlights the importance of GIS methods in demography. Entwisle spoke to the importance of examining

[3] Thiessen polygons construct a "tessellation" of a space or region populated by a set of points located within the region. Each Thiessen polygon encloses that section of the space that lies closest to a particular point.

[4] Network analyses often involve examining spatial relationships on a non-geographic planar coordinate system. Though beyond the scope of this chapter, connections between conventional social networks and spatial networks are a vibrant although emergent specialty. See, for instance, de Smith et al. (2007) for an overview and Okabe et al. (2008) for a software implementation, SANET, for the analysis of spatial networks.

[5] The UCR data are "surveys" of administrative records from cooperating policing agencies.

demographic data in spatial terms. This underlines the increased emphasis on spatial tools in collecting, examining, and analyzing survey data. The inclusion of a chapter on GIS in this *Handbook* is another sign of this.

6.3 Important GIS Terminology

Understanding a few central concepts behind GIS and geographic data themselves (see O'Sullivan and Unwin 2003 for an exposition) facilitates the use of GIS with survey data. Spatially referencing survey data to a geographic location involves linking the data to a *spatial coverage.*[6] A spatial coverage is a digitized version of some element of the geographic landscape. Linking individual-level survey data to a spatial coverage involves 'geo-coding' of individuals to a specific location on the map, using either a physical street address or latitude and longitude coordinates. Assigning a latitude-longitude coordinate to an obtained physical street address is more commonly known as geocoding (see Steinberg and Steinberg 2005; Schuyler et al. 2005).

Three fundamental types of spatial coverages are *points*, *lines*, or *polygons* (see O'Sullivan and Unwin 2003). A *point* denotes a specific geographic locale on a map defined by latitude and longitude (or, alternatively, simple X, Y) coordinates. Points often specify the location of an event or place of interest, such as a household, a committed crime, or a bus stop. *Lines* illustrate features such as roads, rivers, and railroad tracks. They can also denote spatial "interactions" in origin-destination form, such as demographic migration streams. In the social sciences, the most widely used spatial coverage form is the *polygon*, which represents some enclosed spatial area. Examples of polygon coverages include areas contained within political or administrative boundaries, such as nations, states, counties, school districts, and census tracts. Polygons can also represent the approximate or known boundaries of areas defined by social mores and norms, such as reputation-based neighborhoods or a known gang territories within a large urban center.[7]

---

[6] The most popular commercial software vendor, ESRI, calls these "shapefiles" but has recently evolved its conception to spatial data warehouses, called geodatabases. Regardless of the spatial software, shapefiles are usually imported. The file format is open and fully described on the ESRI website (esri.com).

[7] As described in detail in previous chapters, an early exemplar in sociology is Charles Galpin's use of maps of socially-defined rural communities based upon his survey data (Galpin 1915). The Chicago School of thought borrowed Galpin's pioneering ideas and methods of the "natural area" visualized through maps and applied them to "the city" (Park 1929a, b: 61–62). Thus, the use of pre-GIS maps with social surveys led to one of the most influential schools of thought in sociology. Other notable early uses of mapping as a way of understanding social phenomena include Charles Booth's (1887) maps of poverty and John Snow's previously-mentioned work (Rosenberg 1962) on the media through which cholera was transmitted in the mid 1800s.

Figure 6.1 illustrates the three types of spatial coverages, showing how subsets of survey respondents may be represented in a GIS system. The upper-left panel marks a survey respondent's location as a *point*. The GIS references this point to a specific location on the relevant geographic landscape and places it in relation to the other survey respondents. To make these locations more understandable, additional information is needed. A mailing address is usually sufficient for individual survey respondents, but other information such as exact latitude and longitude coordinates can be substituted (see below for more details on the actual procedure).

The upper right panel of Fig. 6.1 illustrates the survey respondent's location in relation to the local street network. The street network exemplifies a *line*-based spatial coverage. Survey respondents are commonly referenced to their spatial locations in the GIS using a street network like this as a spatial coverage. Finally, the lower-right hand panel locates the survey respondents in relation to census tracts, illustrating polygons demarcating areas as a spatial coverage.

For the U.S., many commonly used spatial boundary files (spatial coverages) can be downloaded from the U.S. Census Bureau's cartographic boundary files page (www.census.gov/geo/www/cob). These pre-processed files are drawn from the Topologically Integrated Geographic Encoding and Referencing (TIGER) database (see Ralston 2004). While raw TIGER files require additional processing to produce GIS coverages, some GIS packages (e.g. ArcGIS) read TIGER data files directly. Available freeware code now extracts and assembles raw TIGER data files into ArcView shapefiles or MapInfo files (see Ralston 2004; also available at www.tnatlas.geog.utk.edu). When this chapter was written, the Census Bureau had just begun to release TIGER files in shapefile form.

A key feature of the GIS coverages in the Census Bureau's TIGER files is their geographic identification structure. This structure allows the joining of geographic identifiers and associated attributes to survey data. The most widely used identification structure is the Federal Information Processing Standards (FIPS).[8] Sometimes, however, files use a separate identification scheme developed by the Census or the U.S. Geological Survey (USGS).[9]

The Census Bureau employs a myriad of highly complex geographies, the relationships among which are not always obvious. Survey researchers must become familiar with both the TIGER documentation (see http://www.census.gov/geo/www/tiger/rd_2ktiger/tgrrd2k.pdf) and the Geographic Area Reference Manual (see http://www.census.gov/geo/www/garm.html) before proceeding. Peters and MacDonald (2004) provide a good introduction.

GIS data can be represented in two formats: *vector* and *raster* (see Fig. 6.2 for an illustration). *Raster* graphics are composed of pixels (as in a digital photograph); *vector* graphics are composed of lines (as in line art). The two panels of Fig. 6.2 contain the exact same location data, represented in vector form in the upper panel

---

[8] The following sections on linking data to the GIS explain FIPS codes in greater detail.

[9] The original TIGER files were based on USGS data.

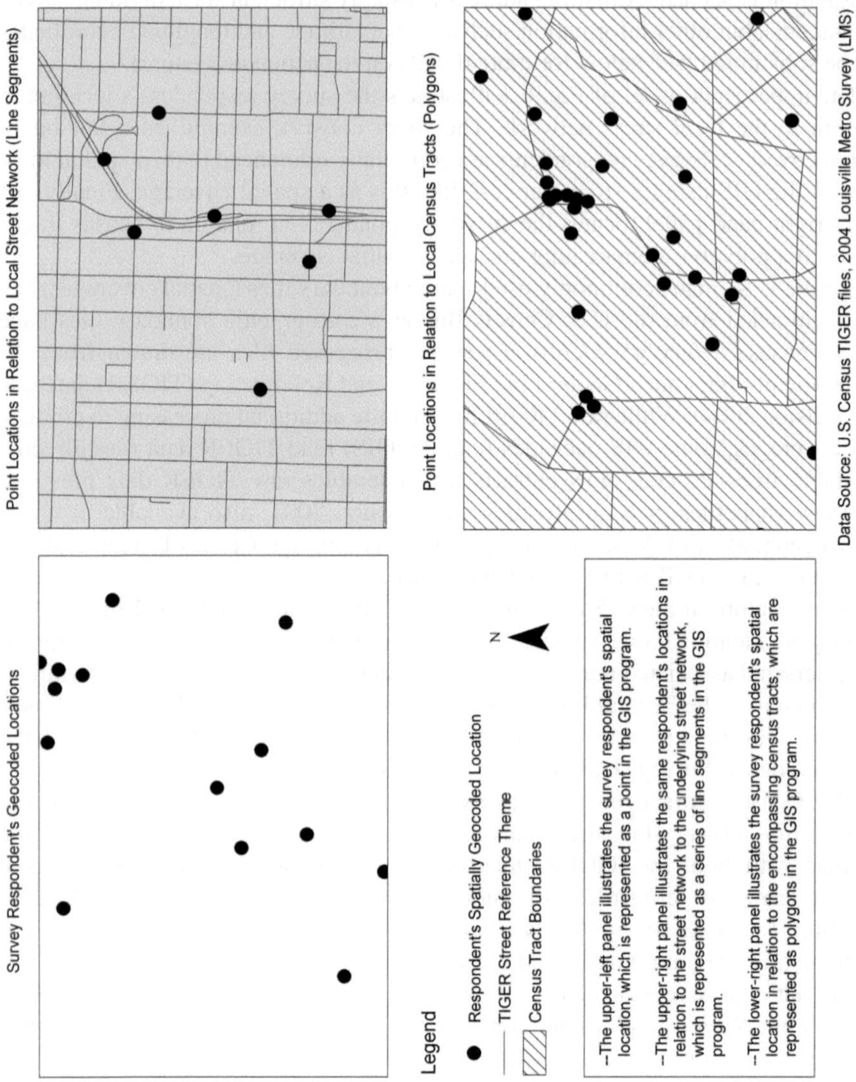

Fig. 6.1 Illustration of geodata in the form of *points*, *lines*, and *polygons* (Howell and Porter 2010)

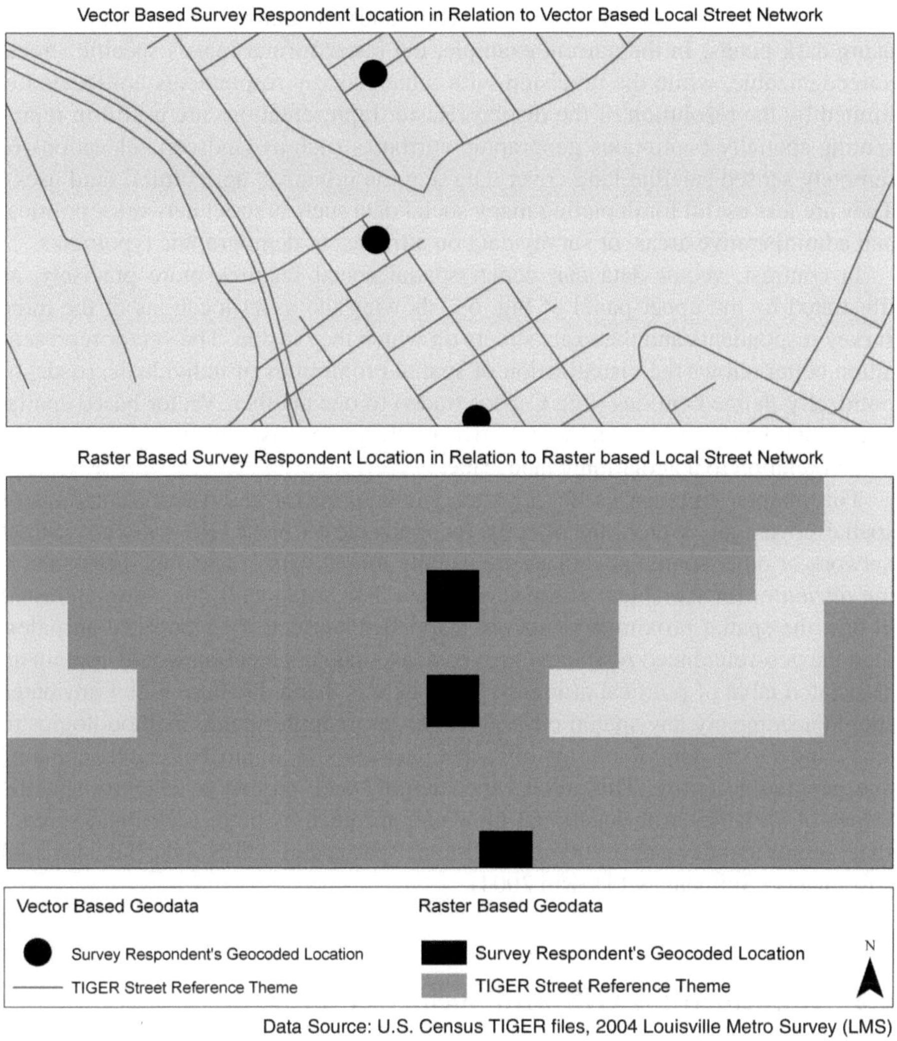

Fig. 6.2 Illustration of geodata represented in a vector and raster based format (Howell and Porter 2010)

and in raster form in the lower one. The raster- and vector-based representations each have advantages and disadvantages; which to use is determined by the needs of the project at hand (see Brown et al. 2005; Boucek and Moran 2004; Ralston 2004).[10]

Pixel-based raster data, in the lower panel, may be associated with a number of geographic characteristics. The shading of pixels identifies different types of land

[10] While this chapter is almost exclusively concerned with the use of vector-based spatial coverages, the raster format is often used in disciplines such as the geosciences, forestry, and urban planning. Examples include such spatial phenomena as pollution levels, forest type, land use/land cover, and other physically identifiable characteristics of the earth.

uses (here, coverage by streets versus vacant land) and denotes survey respondents using dark pixels. In the current example, the raster format makes specific streets unrecognizable, while the precision with which survey respondents are located is limited by the resolution of the display. Raster representations are useful in representing spatially continuous geographic attributes such as landscape elevations or remotely sensed satellite land cover data (e.g. on urban vs. agricultural land uses). They are less useful for depicting many social data such as street networks, political and administrative areas, or survey data on attitudes or demographic typologies.

In contrast, vector data can depict spatial social features more precisely, as illustrated by the upper panel of Fig. 6.2 showing the exact locations of the three survey respondents and the exact streets on which they reside. The vector representation better allows the visualization of spatial proximities of individuals, roads, or politically-defined entities (i.e., Census tracts) to one another. Vector-based spatial representations provide a more intuitive platform for visually examining survey data in relation to a geographic landscape.

This chapter emphasizes linking survey data to vector-based representations of spatial coverages, which are directly recognizable to most GIS software. Street network or other spatial coverages are usually joined with at least one file containing *attributes* (or variables) of the survey data. For instance, if one were interested in how the spatial proximity of survey respondents affects their political attitudes, then the geo-referenced *point* coverage of a respondent's location would contain an associated table of political attitudes.[11] Key here is that a database file of attributes should accompany any spatial coverage in order to apply spatial methodologies to analyzing survey data. Each row of such a database file, or attribute table, contains one geographic entity. This allows the merging and allocation of entity-specific values of variables with locational information, much as more traditional spreadsheet setups identify each entity in geographic space using a GIS and a linked spatial coverage (O'Sullivan and Unwin 2004).

6.4 Implementing GIS with Social Survey Data

6.4.1 Linking Primary Survey Data to a GIS

The most fundamental step in linking survey data to GIS involves the geocoding of micro-data to respondent street addresses. This requires that the survey instrument obtain as much address information as possible—often including postal codes, to avoid problems posed by duplicate street names. This of course raises human subject confidentiality issues, which we address later in this chapter. With the address data,

[11] These associated tables are most often in database format. They can be directly linked to spatial coverages in the GIS program.

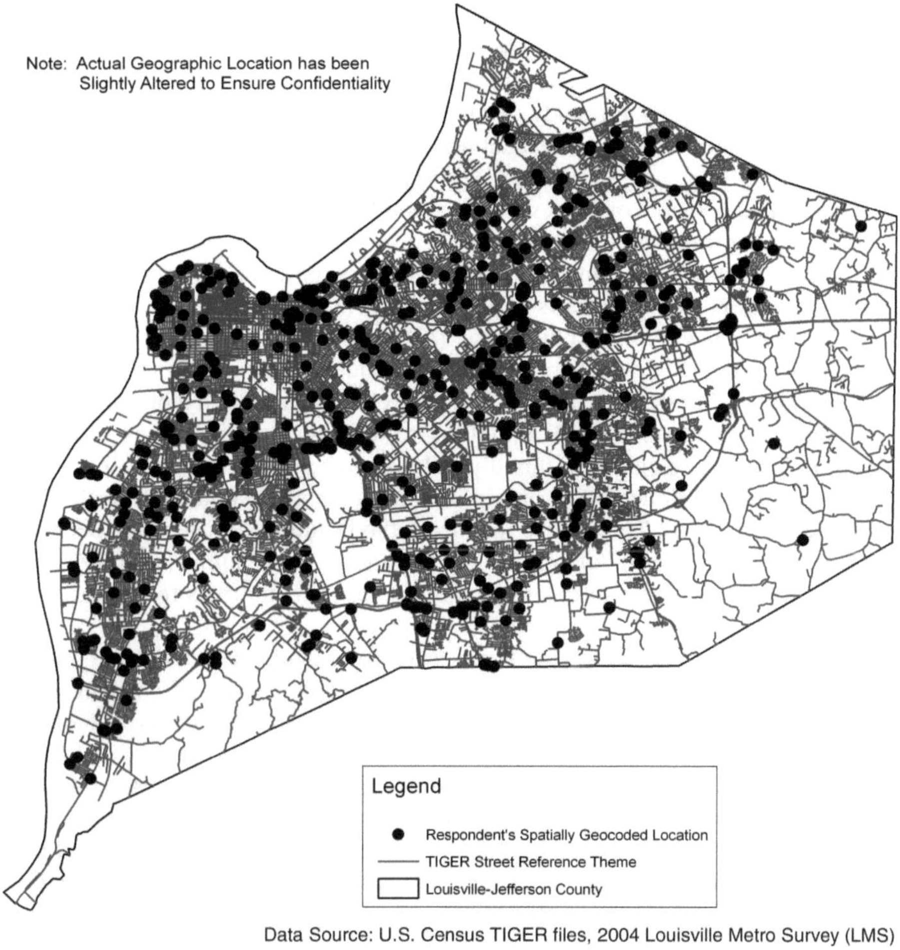

Fig. 6.3 Respondents spatial location via address geocoding, 2004, Louisville-Jefferson County, KY (Howell and Porter 2010)

respondents can be matched to a digitized street shapefile using a defined geocoding reference tool.[12]

Figure 6.3 gives an example result of the geocoding process. It locates 733 respondents in the Louisville Metro Survey (LMS) in relation to a digitized street file. Respondent locations are represented as black *points*, the street network reference file is represented using the lighter gray *lines*, and the Louisville-Jefferson

[12] While the technical aspects of developing a geocoding service are outside the scope of this chapter, we refer the reader to Ormsby et al. (2001) for a "how to" using the popular ArcGIS software created and distributed by ESRI. We provide a cursory overview of the subject later in this section.

County border is represented by the black encompassing *polygon.* The exact locations of respondents were slightly altered in order to preserve their confidentiality when displaying both the respondents and the street-level shapefile in the same figure (see Porter 2004 for study details). Otherwise it would be easy to identify individuals in the less dense areas along the southern and eastern portions of the county.

One common source of address information for geocoding respondent locations is the sampling frame used by a survey. Mail surveys require respondent addresses for delivering the survey and any compensation promised. Likewise, household samples for in-person interviews require that the interviewer have a street address before an interview. As noted by Harter (2010), sampling frames for such surveys can be based on U.S. Postal Service Delivery Sequence Files.

Obtaining address information is more challenging in telephone surveys. Telephone respondents may be unwilling to disclose such identifying information. For landline telephones, this difficulty can usually be resolved by using a reverse telephone directory to obtain a street address (Schootman et al. 2004; Schuyler et al. 2005). The reverse lookup method cannot address problems posed by the increasing number of cellular phones and unlisted numbers, however, for which publicly available directories linking telephone numbers to addresses are lacking.

New trends in survey research are sure to provide researchers hoping to incorporate spatial components into their studies with a series of challenges and obstacles. These trends include the above mentioned increasing preference for cellular over land line phones as well as mounting interest in the usage of web-based survey tools. The use of cellular phones provides a new challenge in that individuals often move, sometimes across vast regions of the country, and do not change telephone numbers. Not only does this create an obstacle to linking a physical address to the phone number via traditional approaches, but it makes linking the individual to geography via area code unreliable. Similar issues surround web-based surveys. With the exception of identifying physical IP addresses, web-based surveys provide very limited information about physical location. They must ask respondents to divulge their street addresses, which many are reluctant to do.

Where appropriate, an alternative to asking respondents in telephone or web surveys for their exact street address is to request their street of residence and the nearest cross street or intersection. This approach does not pinpoint an exact location, but it does place respondents close enough to their homes to examine ecological-level outcomes that may be of interest. Recent work has improved this process by developing better digitized street networks and making reverse address look-up databases more comprehensive (Schootman et al. 2004; Wu et al. 2005).

When the survey data are at the individual level and no street address information is available, some geographic coordinates must be obtained manually. This information is most often obtained using geographic positioning systems (GPS) technology that provides actual latitude and longitude coordinates. This method is primarily used to identify landmarks in more natural resource driven or land-based surveys, but recent improvements in the size and accessibility of GPS technology—coupled

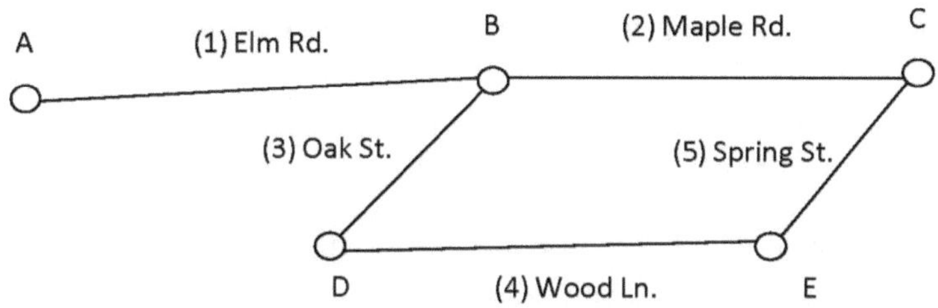

SHAPE	ID	FNODE	TNODE	NAME	TYPE
Polyline	1	A	B	Elm	Rd
Polyline	2	B	C	Maple	Rd
Polyline	3	B	D	Oak	St
Polyline	4	D	E	Wood	Ln
Polyline	5	C	E	Spring	Rd

Fig. 6.4 Illustration of underlying street network shapes and associated database tables (Howell and Porter 2010)

with reduced costs—allow interviewers to gather such information using hand-held devices during face-to-face interviews at a respondent's home (or other location).

Geocoding respondent addresses to their specific spatial locations makes use of the obtained address data and a geographic reference shapefile (i.e. the street network). The most popular geo-referencing database is the street file from the TIGER files maintained and distributed by the U.S. Census Bureau. Figure 6.3 displayed approximate respondent locations in relation to a TIGER street file after it was processed and converted to a geographic shapefile.

Underlying these street files are a series of connecting lines and nodes such as those shown in Fig. 6.4. These allow GIS procedures to locate individuals as accurately as possible. Line segments in Fig. 6.4 represent sections of streets, while nodes represent intersections of sections of streets. An associated reference file contains tables denoting the connecting nodes, address range, and direction of each street segment.

Figure 6.4 depicts the underlying structure of the street network reference theme. Each stretch of street, identified by number and street name in the figure, runs from one node to another, as demarcated by the letters 'A' – 'E'. It also has an associated length and address range (not shown in the figure but given in additional columns of the table in the bottom panel). The GIS referencing tool uses the address information to locate an individual respondent along the stretch, between the nodes. It first identifies an interim street location for an address and then uses ranges of house numbers to interpolate the address's street number along the relevant street segment and obtain a final latitude-longitude result. This results in a series of points representing the location of each survey respondent on a stretch of street, based on her or his physical street address. The address geocoding produces a shapefile with a spatially referenced set of points (representing all survey respondents) and an associated attribute table containing all survey responses for each specific respondent.

6.4.2 Linking Aggregate Secondary Survey Data to GIS

The previous section covered the linkage of micro survey data to *points* in a GIS. This one discusses linking aggregate survey data to *polygon* coverages. The process of attaching survey data to a GIS and associating them with the appropriate geographies is often called "joining." To allow joining, the external survey data and the associated geographic shapefile must have a matching identification structure. This simple requirement often constitutes a pivotal stumbling block. Survey researchers must become familiar with the relevant identification coding schemes for Census geographies before attempting to join attribute and spatial data (good references are Peters and MacDonald 2005; Ralston 2004; Steinberg and Steinberg 2005). If available external data sets for the specific geography of interest, such as a county-level file, use a given identification structure, it is essential to use the same structure when coding locations of survey respondents.

The U.S. Census Bureau mostly uses the FIPS codes created by the federal standards-setting agency, the National Institute of Standards and Technology (NIST), to identify geographic entities in public-use data releases. If a researcher wishes to associate county-level data with individual survey data records, perhaps for a multi-level modeling exercise, the demographic and geographic data could be obtained directly from the Census Bureau in the form of a vector-based polygon file of counties in the U.S. and a database file containing all pertinent variables at the county level. In this case, of course, the survey must obtain information on the county where the respondent resides.

A county-level FIPS code consists of five digits, the first two representing the state and the second three the county. For example, Jefferson County, KY is coded as 21111: the state code for Kentucky is 21 while the county code for Jefferson County is 111. The data record for each survey subject must include the corresponding FIPS code for his or her county location, so that the "external table" of county data can be joined to the survey data.

The FIPS identification structure is the most common linking tool for joining individual-level survey and aggregate-level contextual data. Additional Census geographies at the sub-county level use the county FIPS code as the root portion of their unique code. Tracts, block groups, and blocks in hierarchical succession utilize the FIPS code for each preceding level of geography as a prefix to their full FIPS identification. Other types of ecological data may be more pertinent to a specific researcher's interests than Census geographies, however. In that case, the researcher must examine the available resources before constructing the survey instrument, to ensure that all necessary linking information is obtained.[13] Many federal, state, and private agencies and organizations are influenced by the Census Bureau's practices,

---

[13] Other variations on this general theme of linking survey data to spatial geographies exist, such as spatial processes that include point-in-polygon procedures which counts of the occurrence of an event within a given ecological unit. More comprehensive treatments available elsewhere discuss these (see O'Sullivan and Unwin 2004; Steinberg and Steinberg 2005).

but many others have their own identification taxonomies (e.g., National Center for Health Statistics, Bureau of Economic Analysis). Survey researchers planning to use GIS with data from these agencies must ensure that their data are geographically coded in a compatible format.

6.4.3 Selecting the Optimal Geography for Linking Survey Data to GIS

When planning to link survey data to a GIS, the data needs for a project—in particular, the level of geography required—must be well understood. Using a sub-optimal level of geography and improper methods can lead to inaccurate findings and incorrect inferences (see Anselin and Cho 2002a, b; King 2002; among others). It has long been known that making inferences about individual behavior based on aggregate data is difficult or misleading (see Robinson 1950; Goodman 1953, 1959 (in Anselin and Tam Cho 2002a)); to do so is to risk committing an "ecological fallacy." The ability to link a survey respondent to a larger context using contextual methodologies can reduce the error by introducing individual-level characteristics (Raudenbush and Bryk 2002). One must nonetheless understand the theoretical basis for selecting a particular ecological unit as a context for the individual.

For instance, in studies of ecological determinants of criminal offending, an ongoing debate concerns what unit of analysis is appropriate for conceptualizing "neighborhoods" (Land 1990; Messner and Anselin 2004; Messner et al. 1999; Hipp 2007; Porter 2008; Porter and Howell 2009). Readily available geographies such as counties are unsatisfactory due to extreme within-unit heterogeneity; smaller administrative geographies such as census tracts may also be unsuitable because their boundaries may have little relation to "true" neighborhoods. Grannis (1998, 2005) has shown that traditional thinking about "neighborhoods" must be revised, calling attention to the importance of "pedestrian streets" in defining neighborhoods that cross arbitrary geophysical boundaries.

A related issue arises when working with entity-level geographies: the modifiable areal unit problem (MAUP). Presenting data at one geographic level of aggregation may dilute variation at lower levels of geography (for a recent discussion see King 2002; Anselin and Tam Cho 2002a, b). We illustrate this in Fig. 6.5 by comparing population density at two different geographic levels, for data from the Louisville Metropolitan Statistical Area (MSA). The left panel presents county-level densities, while the right one presents them at the Census block level. These data were obtained from the U.S. Census Bureau's website and pertain to counties and blocks from the decennial Census 2000 survey. The county and block shapefiles were obtained from the Census Bureau's cartographic boundary site.[14]

[14] The URL for the Census Bureau's website is www.census.gov. Geodata can be obtained directly from its cartographic boundaries page http://www.census.gov/geo/www/cob/.

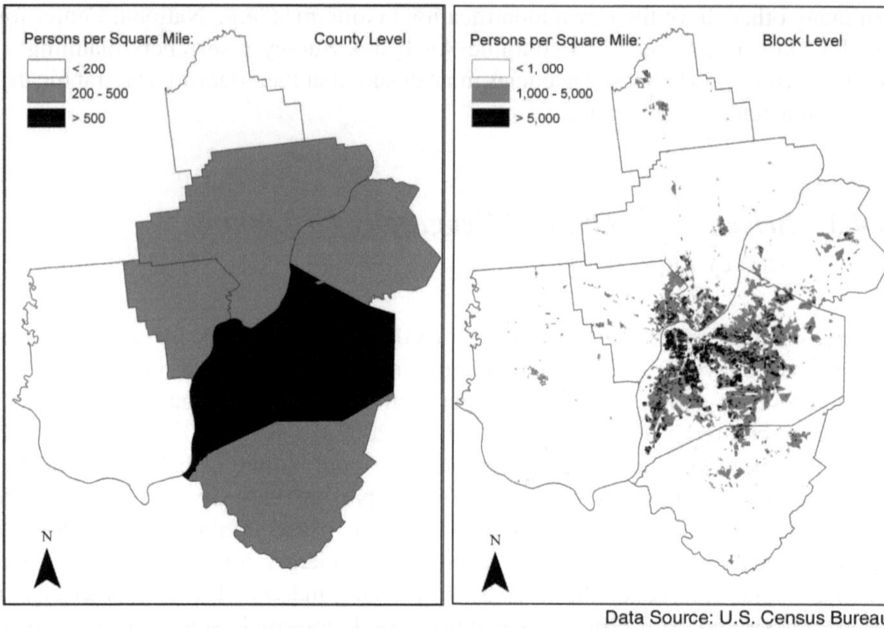

Fig. 6.5 Illustration of within-unit heterogeneity, Louisville, KY-IN Metro Area (Howell and Porter 2010)

The left panel of the figure shows that there is *some* variation in the population density of counties in the Louisville, KY-IN metropolitan area. Its legend ranges from less than 200 to more than 500 people per square mile. Further examination indicates that the county-level aggregation may be less than optimal because it obscures substantial heterogeneity within counties that may be meaningful. The right panel of the figure depicts variation in population density by census block within the same MSA. Its legend ranges from less than a 1,000 people to greater than 5,000 people per square mile. Most counties contain blocks spanning the full range of block-level variation in density. Most of the medium-density counties and much of the highest-density county are as sparsely populated as the low-density counties are.

Clearly inferences based on sub-optimal levels of geography can be inaccurate. For instance, if one were to predict individual-level survey data on behaviors driven by population density using the density measures for the ecological units displayed in Fig. 6.6, findings based on the county-level geography would be much less reliable than those using the block-level attributes. In other research projects concerned with characteristics of cities, towns, or small communities, a place-level geography might be more appropriate.

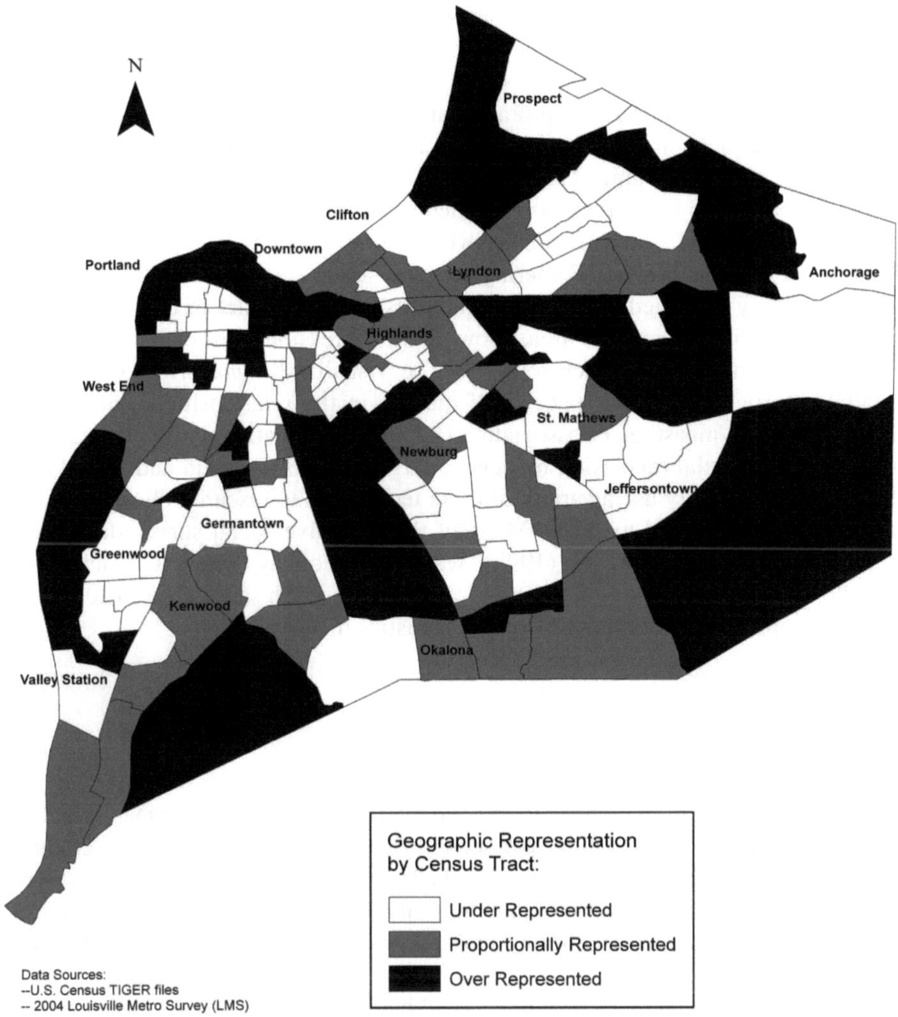

Fig. 6.6 Louisville metro survey geographic representation by census tract, Louisville-Jefferson County (Howell and Porter 2010)

6.5 Moving Beyond Visualization

The increased use of GIS and remote sensing in the social sciences (e.g. Goodchild and Janelle 2004a, b), together with enhanced computing power, has promoted use of high-quality images that display a multitude of information in spatial terms. While this chapter focuses on the linkage of geographical and survey data and their visual interpretation using GIS systems, other spatially-centered statistical methodologies have been developed and employed by researchers in many academic disciplines. Spatial statistical methods allow spatial analysis of social data to move beyond visual inspection and exploratory analysis to fit probability-based models of the role space plays in social relationships. In social science applications, residuals from imperfect models almost always exhibit geographic patterns which can lead to revised models and theoretical improvements (Goodchild and Janelle 2004a, b).[15]

Hence, spatial statistics provide a powerful complement to GIS as a visualization tool. For example, Stewart Fotheringham and associates (2002) have developed geographically weighted regression, which fits a linear model to units of analysis within a specified geographic area, allowing the regression coefficients to vary over areal units. These results themselves can be visualized *a posteriori* to illustrate spatial heterogeneity: how space itself matters in the relationships. Luc Anselin developed local measures of association (i.e. the LISA statistic) and pioneered the estimation of spatial regression models on large datasets (Anselin and Bera 1998). Scholars increasingly use a variety of these techniques to research phenomena that are theoretically linked to space. Many of these involve social survey data in one format or another. See Cressie (1993), Waller and Gotway (2004), Goodchild and Janelle (2004a, b), Lawson (2009), and Raudenbush and Bryk (2002) for introductions to and overviews of spatial statistics methods.

6.6 Chapter Summary

The most significant challenges to the "spatialization" of survey research include the limited attention to teaching spatial analysis methods in social science graduate programs (outside of geography) and the lack of "best practices" standards for protecting human subjects in surveys including spatial data. Meeting both challenges is important if the design, execution, and analysis of micro-level survey data is to embrace spatial thinking.

Survey research methods training should begin to include some training in GIS and spatial data management. When the second author proposed a course on spatial analysis of social data for the Ph.D. program in his Department of Sociology in 1996, the graduate curriculum committee responded that spatial analysis "was not sociology." He argued successfully that neither was the goodness-of-fit chi squared test when Karl Pearson first developed it. During the past decade or so, the social sciences have begun to make spatial analysis methods "theirs" as they once did with cross-tabulation, the Pearson chi-square test, and linear regression. The specialized field of survey research could do well by following this general example.

---

[15] For a more technical treatments, see Anselin (1990), Anselin and Cho (2002a, b), and King (2002), who examine the impact that uncontrolled spatial dependence in data has on the production of unreliable and biased coefficient estimates.

Chapter 7
Spatial Concepts and Their Application to Geo-Sociology

7.1 Introduction

The preceding Chap. 6 discusses ways to make link social data to GIS in order to understand potential ecological impacts that may vary across geographic space. However, the descriptive example analysis does not allow for the testing of ecologically relevant theories of social behavior and group processes. The groundwork of these theories was discussed in the first section of this book (i.e. Chaps. 2, 3, 4, 5). These foundations have been largely geared towards the understanding of social organization along with the development of 'places' or socially-defined localities. Some of these theoretical approaches have focused on the economic efficiency of location attainment, others have focused on the made social adaptation and development their main focus, and still others have focused primarily on the relationship between man and nature in a more physiographic sense. However, all have in common a single interest in the spatial relationships that exist across geographic space, both in regards to the growth of the 'place' and in regards to the proximity of social functions in reference to one another and the larger community.

In this chapter we focus on a few of the primary spatial concepts that link individuals and their aggregates to one another in space. We introduce some of the popular statistical methods for taking these spatial relationships into account in the sociological research. Here we move beyond the description of relationship presented in the preceding chapter to a discussion of methods aimed at linking theory and data through explanatory modeling procedures based on probabilistic statistics. Following this chapter, we discuss these conceptual and methodological introductions with a series of examples, using empirical tests, to highlight their usefulness.

J.R. Porter and F.M. Howell, *Geographical Sociology: Theoretical Foundations and Methodological Applications in the Sociology of Location*, GeoJournal Library 105, DOI 10.1007/978-94-007-3849-2_7,

7.2 Understanding Spatial Concepts in Sociological Research

Underlying the core interests of geo-sociological analysis is the accounting for spatial relationships in sociological research. In particular, sociology as a discipline is particularly interested in the effect social structure, its changes, and its impact on human behavior. Given this interest, sociologists have long been focused on the relationship of the individuals, and groups of individuals, to specific ecological contexts. These relationships can be conceptualized through a handful of spatial concepts, such as containment, proximity, adjacency, and intersection. While this list is not exhaustive, these relationships make up the most common forms of spatial relationships and currently the first three – containment, proximity, adjacency – can be directly examined via the development of sophisticated statistical analyses.

Specifically, the concept of *containment* is one in which the placement of an individual, or group of individuals, within an enclosed boundary is expected to result in orientations that are quite different from individuals that are contained within a different enclosed boundary. It is important to note that we do not necessarily mean and enclosed boundary in the same sense as a wall or fence, but instead in the sense of a city, county, state, or national border. In each case, individuals develop a set of characteristics that are expected to vary in relation to individuals in a different unit.

Similarly, the second spatial concept, the idea of *proximity*, suggests that the closer these individuals are to one another, the more alike they are going to be. Thus, and individual on a border between two units may just as easily take on characteristics of both. However, on average the units themselves are expected to be made up of divergent characteristics.

The characteristics at the unit level are directly related to the third, and final, spatial concept that we will cover in these next two chapters, *adjacency*. When sociologists consider the spatial relationship of adjacency they are most directly concerning with the sharing of borders. Thus, while proximity may relate to an individual's spatial relationship to another person or to a place, adjacency relates directly to the contiguous nature of spatial 'neighbors' in a geographic context.

For instance, the U.S. and Canada would be considered spatially adjacent to one another, while the proximity of the U.S. to Mexico (also adjacent neighbors) is likely to result in more shared characteristics between the two nations than Mexico and Canada, which are much less proximate. Finally, the containment of individuals in the cities of Los Angeles and New York is expected to result in much more city-specific homogenous populations, along many lines of sociological interest, in comparison to the examination of a pooled population from both cities. Of course there are many reasons for these differences, some structural, some cultural, and some related to historical circumstance. We will explore some of these general patterns of spatial differentiation in the following sections.

7.3 Linking Spatial Concepts to Statistical Analyses

In recent decades, technological advances in desktop computing power have greatly improved both the accessibility and opportunity for the development of many software packages that now allow sociologists to test the spatial relationships identified above. Of course, *the first step in the development of any spatially-centered research project is to understand the type of relationship that the research is interested in investigating.* Is it an issue of spatial containment, proximity, or adjacency? If so, there are specific tools that have been developed to handle each of them, all of which are increasing in popularity among published journal articles in the field. These methods are quantitative in nature for the most part, and range from the simple description of populations via a color-based choropleth map, the univariate identification of hot-spots and significant spatial clusters of demographic and social characteristics, and the multivariate spatial weighting of explanatory and predictive modeling. We urge readers to resist choosing the software platform first and tinker with the various options to determine the theoretical question.[1]

7.3.1 Investigating Issues of Containment Through Hierarchical Linear Modeling

From the structural point of view, the most fundamental feature of containment is the idea of *closure*. Closure should be thought of spatially as the nesting of populations or groups within a particular boundary (i.e. spatial unit). This may be a neighborhood, city, state, nation, etc. Furthermore, these "closed" units may be isolated (such as in the case of islands) or may be contiguous (such as in the case of U.S. Counties on the mainland). Depending on the unit of analysis, sociologists may encounter either isolated or contiguous units. Often behavioral and social data have such properties, which require further attention in the modeling procedures used in traditional quantitative applications. Within hierarchical linear models (HLM), each of these spatial units becomes a delineator for a sub-group that lies within its boundaries. Based on this delineation, explanatory and predictive modeling approaches explicitly take in to account the structural relations occurring at that level and any residual variability at that same level of analysis (Raudenbush and Bryk 2002).

[1] For many years, the second author has heard the old standby in statistical analysis, "Oh, I just use SAS™ for everything!" It may indeed be that such programs can "do" the appropriate technique once the conceptual work has been done but it shouldn't be done in reverse. The same applies to other popular software platforms, such as STATA™, SPSS™, or ArcGIS™, just to name a few. That is, researchers should choose the tool and software implementation after the analytical procedure is appropriately identified.

For the purposes of this chapter and book, we will focus on the two-level HLM, however, the current HLM™ package does allow for up to four-level models. Such may be appropriate for an analysis of individuals, within counties, within states for example. In such a case, one might suspect that local structural variations exist across counties, but they are further impacted by state-level legislative differences. In this case, the appropriate HLM would be one that models the structural variations at both ecological levels as individuals are nested both within counties and states, but also counties are nested within states. Here we focus once again on the approach that individuals, or groups, are nested within a set of larger ecological units. Thus, we present a traditional two-level model by drawing heavily on the HLM6 handbook (see Raudenbush and Bryk 2002 for any points of clarification throughout this discussion).

A general HLM analysis happens over several nested steps and therefore is multi-phasic (Raudenbush and Bryk 2002). There are four basic steps to each analysis, all of which will be explained in this section. Estimating each of the four models using the HLM6 software will yield coefficients and standard errors that adjust for both level-1 and level-2 variation as opposed to a single universal variance component applied to more traditional analytic techniques (Raudenbush and Bryk 2002). The four models, in the order they are traditionally presented, are (1) the One-Way ANOVA model, (2) the Regression with Means-as-Outcomes model, (3) the Random-Coefficient model, and (4) the Intercepts- and Slopes-as-Outcomes model. Each is outlined in greater detail below. For the following presentation, a theoretical case of the occurrence of individual acts of crime (level 1) within a specific set of neighborhoods (level 2) will be employed.

Step 1: HLM One-Way ANOVA Model. The first step to understanding the potential effect of variations in neighborhood conditions on spatial variations in crime, within the HLM context, is to estimate a One-Way ANOVA Model in order to obtain important and useful preliminary information concerning the level-2 grand mean and the amount of variation in the response variable accounted for within and between discipline types. Specifically, this procedure results in the ability of the researcher to gain insight into the magnitude of the effect associated with ecological differences as opposed to individual differences. This procedure results in the following estimation:

$$Y_{ij} = \gamma_{00} + u_{0j} + r_{ij}$$

The equations specification predicts the individual level occurrence of a crime being committed (Y_{ij}) that is equal to the overall grand mean level of crime across the entire population (γ_{00}) plus random error at both the neighborhood (u_{0j}) and individual level (r_{ij}). Within the HLM context the inclusion of both neighborhood and individual level error terms allows for the control of standard error issues associated with the use of a single error term in more traditional statistical techniques (i.e. regression approaches that do not explicitly model individuals within ecological units as sub-groups). The results of this model will provide us with fixed effects in the form of the grand mean (as described above), but more importantly the HLM results will provide us with random effects in the form of the estimated individual level variance component (σ^2) and the estimated neighborhood level variance component (γ_{00}). These variance components can be used to calculate the initial proportion of variation between neighborhoods through the following equation:

$$\rho = \tau_{00} / (\tau_{00} + \sigma^2)$$

Here the amount of variation accounted for between neighborhoods (τ_{00}) is divided by the total amount of variation ($\tau_{00} + \sigma^2$) resulting in an initial estimate of potential neighborhood level contextual effects.

Step 2: HLM Regression with Means-as-Outcomes Model. The next step in the HLM analytic process is to estimate the Regression with Means-as-Outcomes model in which the individual level occurrence of crime is viewed as varying around neighborhood means that are further predicted by any set of theoretically related neighborhood (level 2) characteristics. In the case of this specification, we assume that the percent of the population in poverty is related to higher rates of crime given the prior documentation of these relationships. This marks the first point at which the difference in the occurrence of crime is examined across the neighborhoods based on neighborhood characteristics, but does not include any level-1 predictor variables in the model (i.e. demographic, cultural, or other characteristics of the individuals committing crime). Although the model is only testing the differences of mean levels of crime by the percent in poverty, both individual level and neighborhood level error terms are again included in the mixed level model since level-2 neighborhood characteristic is predicting the level-1 individual crime variable. The HLM Regression with Means-as-Outcomes Model is specified as:

$$Y_{ij} = \gamma_{00} + \gamma_{01}[\% \text{ Poverty}] + u_{0j} + r_{ij}$$

Here the individual occurrence of a crime being committed (Y_{ij}) is equal to the overall grand mean level of crime (γ_{00}) plus random error at both the neighborhood (u_{0j}) and individual level (r_{ij}). However, unlike the One-Way ANOVA model, the fixed effect of the percent in poverty is taken into account ($\gamma_{01}[\% \text{ Poverty}]$) in order to identify potential effects across neighborhoods by varying rates of poverty and in relation to variations in the occurrence of crime level. The resulting parameter will provide an estimate of the independent effect of the percent in poverty on the occurrence of crime.

Of course, it will be important to use the resulting variance component information from the HLM in order to better understand the effect of the inclusion of the level 2% in poverty as a predictor in the model. In this case, the variance component associated with contextual level is a residual (τ_{00}) left over from the variance accounted for by percent in poverty. This residual can be used along with the baseline level-2 variance from the earlier ANOVA model in order to develop an index of the proportion reduction in variation. The proportion will tell us the amount of the potentially explainable level-2 variation that can be accounted for by variations of the percent in poverty at the neighborhood level and is specified as:

$$\rho = \left[\tau_{00}[\text{ANOVA}] - \tau_{00}[\% \text{ Poverty}]\right] / \tau_{00}[\text{ANOVA}]$$

Here the proportional reduction in variance (ρ) is equal to the total neighborhood level variance (τ_{00}[ANOVA]) minus the residual variance left over from the level-2 discipline type Means as Outcomes Regression Model (τ_{00}[% Poverty]) divided by the total neighborhood level variance as a baseline. Again, the results for this sample analysis would indicate the amount of variation in the occurrence of crime explained solely by variations in the percent of the neighborhoods in poverty.

Step 3: HLM Random Coefficients Model. After the effect of the percent in poverty is examined, next the specification of the HLM Random Coefficients Model will allow researchers to understand the effect of individual level characteristics on the individual level occurrence of crime. For this example, we take a structural approach and are interested in potential variations across racial categories. This model does not include any neighborhood characteristics, but within the HLM context this is different from a traditional single level model, as the hierarchical context of the data remains an important part of the estimation process. Within the HLM context, the level-1 effects will be modeled independently for each of the neighborhoods and then the average intercept and slope are reported, thus continuing to allow for the fact that the these crimes occurred in a nested multi-level context to be included in the modeling process (Raudenbush and Bryk 2002). Here, we will test the level-1 effects of race as predictive of variations in the occurrence of crime. The Random Coefficients Model can be specified as:

$$Y_{ij} = \gamma_{00} + \gamma_{10}[\text{Race}] + u_{1j}[\text{Race}] + u_{0j} + r_{ij}$$

Here the individual level occurrence of crime (Y_{ij}) is equal to the overall grand mean level of crime (γ_{00}) plus the average regression slope for the race of the offender on the response variable (γ_{10}[Race]) plus the unique effect of the discipline on the associated race slope (u_{1j} [Race]). Also the random error at both the neighborhood (u_{0j}) and individual level (r_{ij}) are again taken into account.

Next we will again use the resulting variance component information to examine the proportion variance explained at level-1 from the following equation:

$$\rho_1 = \left(\sigma^2[\text{ANOVA}] - \left(\sigma^2[\text{Race}]\right)\right) / \sigma^2[\text{ANOVA}]$$

Here the proportion variance explained at level-1 (ρ_1) is equal to the original level-1 variance identified in the one-way ANOVA model (σ^2[ANOVA]) minus the variance from the level-1 model containing race (σ^2 [Race]) divided by the original level-1 variance from the ANOVA model (σ^2[ANOVA]). This result will report the reduction in variance in crime accounted for by the effects of taking the offender's race into account.

Step 4: HLM Intercepts- and Slopes-as-Outcomes Model. Now that the variability in regression equations across the occurrence of crime has been estimated at both the individual and neighborhood levels, here we expand the level-1 model to include level-2 main effects and any potential cross products. This model illustrates the first point at which the interacting effects of race and poverty are introduced into the analysis. For this example analysis, the full Intercepts- and Slopes-as-Outcomes model is specified as:

$$Y_{ij} = \gamma_{00} + \gamma_{01}\left[\%\text{ Poverty}\right] + \gamma_{10}\left[\text{Race}\right] + u_{1j}\left[\text{Race}\right] \\ + \gamma_{11}\left[\%\text{ Poverty}\right]*\left[\text{Race}\right] + u_{0j} + r_{ij}$$

Here the individual level occurrence of crime (Y_{ij}) is equal to the overall grand mean level of crime ($_{00}$) plus the main effects for the percent of the neighborhood in poverty ($_{01}$[% Poverty]), race of the offender ($_{10}$[Race]), and the cross-level interaction effects of the percent in poverty on race (γ_{11}[% Poverty]*[Race]). This cross-level interaction is particularly interesting as its results estimate the potential 'intensification' or 'relaxation' effects associated with neighborhood levels of poverty on the relationship between race and the committing of crime. Again taking a structural viewpoint, one might hypothesize that the reason that crime is disproportionately committed by minorities is the social fact that they make up a larger proportion of the population in poverty. By examining potential neighborhood effects in this example, we can test to see if living in a neighborhood of high poverty increases (intensifies) or decreases (relaxes) the relationship of being of minority status in regards to the incidence of crime. Finally, the equation also includes the unique error associated with the level-1 slopes for race (u_{1j} [Race]) and random error at both the discipline (u_{0j}) and student levels (r_{ij}).

As with all previous models, we will again compute the proportion in variation explained for each of the random coefficients via the following equation:

$$\rho_2 = \left[\tau_{qq}\left[\text{random coefficients}\right] - \tau_{qq}\left[\text{fitted model}\right]\right] / \tau_{qq}\left[\text{random coefficients}\right]$$

Here the variance component for the percent in poverty mean intercept and race for this full fitted model (τ_{qq}[fitted model]) will all be subtracted from and divided by the variance component for the previous random coefficients model (τ_{qq}[random coefficients]). The final results will indicate the total amount of variation in the individual level variance of race as a predictor of the committing of crime that is directly related to the neighborhood level percent in poverty.

Ultimately, the mixed-level modeling strategy discussed here is very useful for understanding the ecological effect of geographic areas in which social phenomena occur (such as crime, unemployment, and so forth). In estimating the series of HLMs presented above, and in relation to one's own research interests, the ability to account for the structural variables that are theoretically related to individual level variations in a specific behavioral or social outcome are possible. Today most of the popular statistical programs used by sociologists allow researchers to control for the effect of level 2 and 3 ecological effects. HLM™ itself is a specialized package for the estimation of such models, but SAS™, STATA™, R, MPLUS™, and other popular packages contain similar methods for such estimation.

7.3.2 *Investigating the Effects of Proximity Through Distance Based Measures of Closeness*

When measuring proximity, the most important indicator is that of relative distance between or among spatial objects. Proximity is a distance based measure of the relationship of two (or more) units across a spatial plane. Sociologists using a geo-sociological approach may be interested in various units accounting for the proximity of persons in relation to one another, populations in relation to one another, or even persons in relation to physiographic features. In relation to the spatial features presented in the previous chapter, points and polygon are most useful in these types of analyses as points most often represent a person's location, a physiographic feature, or the center of a containment area and polygons represent aggregate catchment areas of populations. *Point pattern analysis* is perhaps the least used form of spatial analysis to this point in sociological research, largely to the lack of spatial location information attached to micro-level data but this is changing in survey research (Howell and Porter 2010). However, many tools and methods have been developed for point pattern analysis, especially in the areas of spatial crime as well as public health research.

In this section we will present information on specific types of point patterns and the underlying processes that help to shape those patterns. Much of the information presented here was gained from Waller and Gotway's (2004) *Applied Spatial Statistics for Public Health Data* and will reference the uneven occurrence of health outcomes in presenting this information. Specifically, we will couch our presentation of these methods in the real-world relationship between the uneven spatial distributions of cancer incidence that has been well documented in the epidemiology literature.

In regards to the patterns that underlie the distribution of point patterns, there are two basic arrangements that are of primary interest to sociologists. First, is perhaps that most interesting and is the *clustered* pattern of points, and the second, which is actually to opposite of the clustered pattern, is the *uniform* or *random* pattern. When patterns are found to be clustered they violate this uniform distribution across the spatial plane and are closely located to one another in terms of distance. In contrast, a uniform or random distribution would meet a probabilistic assumption of even spread across the spatial plane. Thus, one might envision an even distribution look similar to the generic plots in Fig. 7.1. From the plots one can see that the distribution of 'x' in the left hand panel is perfectly random given the even spatial distribution of all points. However, the spatial distribution of the 'x' in the plot to the right hand side of the figure show a spatially uneven pattern of distribution. If we were to assume that these 'x' each demarcated the spatial location of an individual with cancer, we would draw tremendously different conclusions from the two patterns of these locations. From the left hand side, we would assume that there was no underlying relationship between the likelihood of having cancer and the individual's proximity to a specific spatial location. However, if we look at the right hand panel,

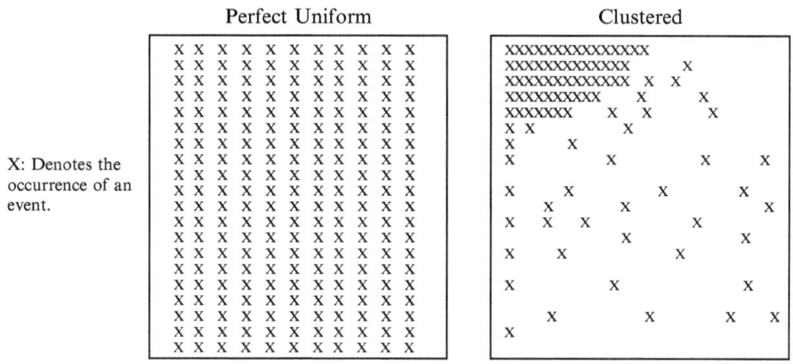

Fig. 7.1 Generic illustration of a perfect uniform and clustered spatial distribution

we see that there is a direct relationship with being in the upper left hand corner of the plot and having a higher incidence of cancer, this may be considered an *event location*. Now, if we move beyond the spatial plane and imagine a similar distribution across a specific geographic space, say the state of Kentucky, we would assume that there is some social or environmental process occurring in the northwest corner of the state that is associated with high rates of cancer.

There are a number of associated spatial point processes that must further be taken into account in order to move towards the identification of spatial clusters, in this hypothetical case, spatial clusters of cancer. First, there are two basic concepts that provide a starting place for the modeling of spatial point processes (Waller and Gotway 2004). These concepts represent the point processes that can be described as *stationary* and *isotropic*. Stationary processes are dependent only on the relative position of each point to one another and not on potential event locations. In addition, isotropic processes are dependent on neither an event location nor orientation, but only on the proximity based distance between points. These represent the simplest type of spatial processes and are not very useful in the identification of underlying causes of, or the identification of an event location, but do provide a framework from with the proximal relationship between points can first be understood.

In order to uncover potential relationships to the geography on which real world social processes occur, we must move beyond the simple proximity based assumptions of the stationary and isotopic point processes to one that takes into account the unit area. Waller and Gotway (2004) present one such process, the Spatial Poisson Process as one such process that allows for the identification of point distributions to be statistically identified as random or clustered. In its most basic form allows for the calculation of an expected number of events per unit through the following calculation:

$$\gamma = N / |A|$$

Here the expected intensity (γ) is equal to the number of points (N) divided by the absolute area ($|A|$). Again, this is the expected intensity and thus represents a perfectly even distribution of points across the spatial plane. For instance if a unit have four incidents of cancer, and under the assumption of a square spatial plane, the expected intensity would be that they would be perfectly spaced and fall within four separate quadrants of the spatial plane. This then represents a completely spatial random (CSR) pattern (Waller and Gotway 2004). Without taking any other geographic characteristics into account, this would represent a case in which there is constant risk to all member of the population based on spatial proximity.

To test potential hypotheses concerning the CSR pattern of a specific point distribution, researchers must also take into account specific issues of population density as it is logical to assume, and a real-world social fact, that incidents of non-infectious and infectious diseases are more prevalent where there are higher levels of density. This is due the fact that at the same probability rate of incident, place with more population will have more incidences. Thus, the identification of a spatial cluster may not always be related to a violation of the assumption of constant risk across the spatial plane.

A number of statistical packages allow for the testing of these relationships. One such software package is GeoDa, version 0.9.1 (see http://geodacenter.asu.edu). GeoDa allows for the estimation of "nearest neighbors" of "distance" based spatial weights. Both are appropriate for different research questions. For instance, recent research on the diffusion of crime from cities and towns to rural areas in a given county found that the most efficient definition was to link each rural area to the three closest cities or towns (see Porter 2010 for more information). Regarding the distance-based weight, research interested in the isolation of rural communities from metropolitan centers may be interested in understanding how far those communities are from urban centers (such as in the food desert research by Blanchard and Matthews 2007). These relationships can also be further examined in the form of a multivariate analysis in which these weights are incorporated into a regression analysis to account for the predictive ability of proximity in explaining variations in a specific outcome variable. Spatially weighted regression models and a method of identifying significant spatial clusters across geographic space are presented in the following section.

7.3.3 Investigating the Effect of Adjacency Through Spatial Regression and Spatial Clustering

The last spatial concept to be illustrated here is the relationship of *adjacency*. The most popular form of measuring adjacency in sociology is the contiguous relationship of places to one another. For instance, the contiguous nature of neighboring counties has been often been found to be significantly lined to the spatial clustering of many social phenomena. Similar to the identification of distance-based spatial clusters, the identification of contiguously based spatial clusters is based on the assumption that spatial clusters of social phenomena can be identified through the examination of statistical models aimed at testing observed distributions against uniform distributions of data across space. Once the identification of a significantly uneven spatial distribution of data is made, predictive analyses can incorporate spatial weights for controls in further analyses.

These analyses are most often undertaken in multi-phase fashion. If it is suspected that spatial clustering in a specific variable exists, the initial descriptive stage involves Exploratory Spatial Data Analysis (ESDA) (see the tutorials at http://geodacenter.asu.edu for further information). Here we present a global indicator of spatial autocorrelation, the Moran's I statistic (Moran 1950). The most common case is that of positive autocorrelation in which the local unit's (*i*) value on a variable of interest is significantly, and positively, correlated with the average neighborhood (*j*) value. Less frequently, negative autocorrelation refers to an instance when a local unit's (*i*) value pertaining to a specific variable is significantly in opposition to the neighborhood's (*j*) average value. The Moran's I is specified in the Following equation:

$$I = \frac{1}{s^2} \frac{\sum_{i=1}^{N} \sum_{j=1}^{N} \omega_{ij} \left(\overline{Y}_i - \overline{Y}\right)\left(\overline{Y}_j - \overline{Y}\right)}{\sum_{i=1}^{N} \sum_{j=1}^{N} \omega_{ij}}$$

In the above equation the measure of spatial dependence is equal to a measure of variation in the area unit specific rate and the overall mean rate (s^2) is multiplied by the neighbor weight indicator (ω_{ij}), times the product of each unit (*i*), minus the overall mean and each neighborhood (*j*), minus the overall mean then divided again by the weight indicator and summed across all units (*i*) and across all neighborhoods (*j*) (Moran 1950; Waller and Gotway 2004). The statistic is very similar to Pearson's Correlation Coefficient in that it measures an association between *N* observed values associated with two random variables, *Xi* and *Yi* (Waller and Gotway 2004). In this case the only difference is replacing the *Xi* variable with the *Yj* neighborhood variable and introducing the weight matrix (ω_{ij}).

This equation produces a statistic in which each unit's (*i*) interaction with another is taken into account and when neighboring units (indicated by a 1 as the ω_{ij}, as opposed to a zero for non-neighboring units) are statistically significant and similar the Moran's I statistic is positive, meaning areas of closer proximity tend to be more alike than those far apart (Waller and Gotway 2004). In this instance the result is spatial clustering. In order to place a significance value on the observed Moran's I statistic, a permutations based tested will be implemented to test the null hypothesis: "No spatial association" or "spatial randomness". The test employs a permutations based approach to test the global index on randomly assigned locations in order to approximate the distribution of the global index under the null assumption (Waller and Gotway 2004). This examination will implement a test of 999 permutations with a reject region equal to a 0.05 significance level, although a smaller number of permutations could be used.

In order to identify statistically significant independent clusters of contiguous "zones" among counties, we use Anselin's Local Indicator of Spatial Association (LISA) statistic (1995), which is based upon the global Moran's I coefficient decomposed into a "local" level, which is the county in this study. This local examination repeats the spatial clustering procedure for each neighborhood, which is equivalent to the reproduction of the procedure *i* times (once for each county accompanied by all of its identified neighbors) (Waller and Gotway 2004). This procedure ultimately produces a categorical outcome based on the relationship of county *i* to the remainder of the counties within the *j*th neighborhood, producing a result that indicates positive spatial clustering (county is significantly like its neighbors), negative spatial clustering (county is significantly unlike its neighbors), or spatial random distribution (county is not significantly like or unlike its neighbors) (Anselin 1995). For the purpose of this paper, the general LISA statistic will be employed. It is specified in the following equation as:

$$I_i = \sum_{j=1}^{N} \omega_{ij} \left(Y_i - \bar{Y}\right)\left(Y_j - \bar{Y}\right)$$

From this equation it is apparent that the random variable (LISA) I_i is equal to the weight indicator multiplied by the product of the local unit value (Y_i), minus the global mean ($\bar{Y}$) and the neighborhood average value of the same variable (Y_j), minus the global mean ($\bar{Y}$). Simply, the LISA value for a given location is equal to the relationship between the two variables of interest (correlation) multiplied by the weight indicator matrix (one if considered a neighbor, zero if otherwise).

Based on the findings from the ESDA in the descriptive analysis, any subsequent explanatory regression analysis should identify the appropriate technique to control for the existence or absence of spatial autocorrelation. In the event of obvious spatial autocorrelation, spatial dependence diagnostics are examined in order to select the appropriate spatially weighted regression model. In order to test for the model specification to control for the identified existence of spatial autocorrelation, OLS models are initially run employing an adjacency weight matrix. If conducted in GeoDa, these preliminary models will return a series of coefficients, via the Lagrange Multiplier Tests, which identify whether the spatial autocorrelation is due to spatial lag or error.

Ultimately, the two most employed forms of spatially weighting regression models entails the introduction of a spatial weight to account for the correlation of error terms or to account for the actual substantive role that adjacency has in the explanation of variations in a specific outcome variable. Introducing a weight for spatial error into a model entails correcting for non-random error term correlation. This is done by adding the spatial weight to the error of the equation; also known as a Simultaneous Autoregressive Model (SAR). Introducing a weight as the spatially lagged version of the dependent variable is known as a Conditional Autoregressive Model (Cressie 1993; Waller and Gotway 2004). Each of these models is estimated as follows:

$$(\text{SAR}) \qquad Y_i = x_{ij}\beta + \varepsilon_{ij} \quad \text{where:} \quad \varepsilon_{ij} = \lambda\omega\varepsilon + \upsilon$$

$$(\text{CAR}) \qquad Y_i = x_{ij}\beta + \rho\sum\nolimits_j \omega_{ij} y_j + \varepsilon_{ij}$$

The equations show that the SAR model consists of a basic regression model with the only weighting taking place in the error term (ε). Here, the correlated error terms are controlled through the introduction of a spatially weighted adjacency vector ($\lambda\omega\varepsilon$) and an added vector of independently and identically distributed (iid) errors (υ). This spatial weight, along with the introduction of the uncorrelated error terms, allows for the regression coefficients to be estimated without violating the assumptions inherent in spatially correlated data (Cressie 2003; Waller and Gotway 2004). It is evident that the CAR model is also similar to a standard linear regression model where the second term is constructed from a predefined n by a spatial weighting matrix, (ω_{ij}), applied to the observed variable, (y_j), together with a spatial autoregression parameter, (ρ), which typically has to be estimated from the data (Cressie 2003; Waller and Gotway 2004).

There are other forms of regression that take spatial concepts into account. For instance, Geographically Weighted Regression (GWR) takes a slightly different approach in that it computes a series of regression models across different sections of the spatial plane (Fotheringham et al. 2002). This is in contrast to the regression models specified above that control for the specification of neighbors through the introduction of a neighborhood weight matrix (1 = neighbor; 0 = not a neighbor). The approaches are similar in that they both control for the geographic proximity of regression cases in reference to one another, but unlike the spatially weighted approaches presented above the GWR approach does not weight the regression equation itself in treating the whole map in a single approach. Instead the GWR approach estimates a series of regression equations for every specified spatial unit. The GWR approach is estimated as:

$$Y_i(u,v) = b_0(u,v) + b_1(u,v)x_1 + e(u,v)$$

Here the only difference between this model and a traditional regression model is the inclusion of a set of spatial parameters meant to define a specific location (u,v). However, the estimation is another story. There are *ith* numbers of spatial locations in any given map based on the area of interest and the theoretical spatial "reach" of independent variables. In essence, there are "different slopes" for "different folks" where "folks" are the spatial units under analysis! While this is an inductive analysis, where the slope estimates for each spatial unit are not specified prior to estimation, it can impact theory-building or modification in important ways. Most dramatically, a sub-region of spatial units can have a positive relationship to a predictor while others have a negative one. The two opposite-signed directions may cancel each other out so that the conventional constant-parameter models might show no association. Or it may be only a small sub-region of spatial units has a significant association in either direction which would be "washed out" by the majority of spatial units which together have a random relationship. GWR is also a good exploratory check against constant-parameter models for the presence of "spatial regimes" of an unknown composition (see Anselin 1988 for details of spatial regime models).

7.4 Conclusion

The continued advancements in computational technology and the increased accessibility of statistical programs with the capability of performing the analyses outlined in this chapter have helped to drive the increased presence of ecological context in sociological research. At their core, these are the approaches which allow sociologists to study the topics that are of most interest to the discipline. For instance, how is it that social organization and the physical environmental surroundings impact individual behaviors? Furthermore, what is it about that relationship that produces these interactions? How does one type of organizational pattern (e.g., population density) affect another (e.g., economic decline)?

Here we have presented three basic spatial concepts and popular approaches to empirically examining the existence of each. We understand that there are many alternative approaches to the ones specified here and do not suggest that these are the most correct or efficient approaches to examining such spatial relationships. However, their application and continued usage in the field has helped to situate them as useful methods for such an examination. In the next chapter we will present and interpret empirical our own empirical tests of these methods.

Chapter 8
Geo-Sociology in Practice

8.1 Introduction

In this chapter we will provide examples of data that has been linked to GIS or appended to ecological characteristics and undertake the analyses previously introduced. Here we present an example of the relationship between poverty, stress levels, and self-rated health within a hierarchical linear model (HLM) context. In this analysis we show that individual level stress negatively effects perceptions of their own health. This health perception is further impacted by the ecological environment in which they live, partly by the neighborhood poverty rate. This represents the issues associated with the containment of individuals within a specific neighborhood context and the effect that ecological containment has in effecting individual level phenomena (i.e. perceived level of health).

Next, we examine the spatial concept of proximity with an analysis of event specific locations in reference to one another. Here we present an example of point pattern analysis using crime data from the Atlanta, GA metropolitan area. In this example analysis, we use the spatial location of homicides in the city of Atlanta from 2004 to 2009 (N = 545). Using that data, we present a simple set of descriptive tools that can be used to help understand the distribution of the homicides. We also incorporate nearest neighbor distance weights in order to identify significant clusters of high levels of homicide over this time period.

Our final example analysis is an examination of the spatial concept of adjacency. In this example, we discuss one of a number of procedures through which a researcher might go about detecting significant spatial relationships before conducting a simple regression analysis. Substantively, we extend the "protection hypothesis" that marriage provides a positive effect in regards to health by examining a potential ecological relationship between the average life-expectancy of counties in the U.S. by the percent of the population that is married.

While all of these represent rather simplistic approaches to the application of spatially-centered analytic techniques, they are intended to highlight some of the

J.R. Porter and F.M. Howell, *Geographical Sociology: Theoretical Foundations and Methodological Applications in the Sociology of Location*, GeoJournal Library 105,
DOI 10.1007/978-94-007-3849-2_8,

more popular approaches to geo-sociological research by discussing these applications with real-world data and measures that are generally of interest in sociology.

8.2 Self-perceived Health, Stress, and Neighborhood Poverty: An HLM Approach

In testing potential issues of containment, it is important to understand the spatial relationships between units of analysis. While not always the case, one of the most popular approaches is to examine the effects of aggregate units of geography on individual-level social and behavioral outcomes. Figure 8.1 presents such a relationship with a set of geocoded respondents from the Panel of American Life Survey (PALS) (see Chap. 6 for more information of geocoding). Once these data are spatialized, they can be further attached to larger units of analysis. In this figure, we have illustrated the relationship between the PALS survey respondents and the census tract boundaries that contain their residencies. For the purpose of this example analysis, these tract boundaries are considered neighborhoods in much the same way that they approximate such areas in U.S. cities. From the figure, one can see that each neighborhood has enclosed within its boundaries a series of survey respondents. In all, the PALS dataset is made up of 2,600 survey respondents across nearly 300 census tracts. Taking an HLM approach allows for the ability to model each individual neighborhood's structure in the relationship between individual level variables, while also allowing for the effects of indicators from the neighborhood level in both main and interaction form.

In this analysis we have used data from the PALS and examined the multilevel relationships between stress, self-perceived health, and neighborhood poverty. At the individual level, the relationships between levels of stress were computed via a ten item index that proved to be reliable for this example analysis (alpha = 0.829). Self-perceived health is a dichotomous measure of a respondents report that their health is only fair or poor (in reference to excellent or good). Finally, the neighborhood poverty measure is an indicator of the percent of the census tract that is classified by the U.S. Census Bureau as being in poverty. The descriptive statistics report that the average stress score is 2.64 (out of 10), 23% of the respondents classify their health as poor or fair, and the mean poverty level across the neighborhoods is about 14%. The modeling strategy is specified so that we are interested in the effects of stress, at the individual level, and the effects of poverty, at the neighborhood level, on one's self-perception of their own health. Furthermore, the modeling process follows the discussion of the technique in Chap. 7 in order to highlight each step in the modeling process.

Step 1: HLM One-Way ANOVA Model. From the results of one-way ANOVA model, we find that a significant proportion of the variance in self-reported 'bad' health measure is accounted for across level to units at the neighborhood level (variance component equal to 0.29, *p-value* <0.001). This initial finding suggests that taking only an individual level (traditional single level) approach to modeling this health

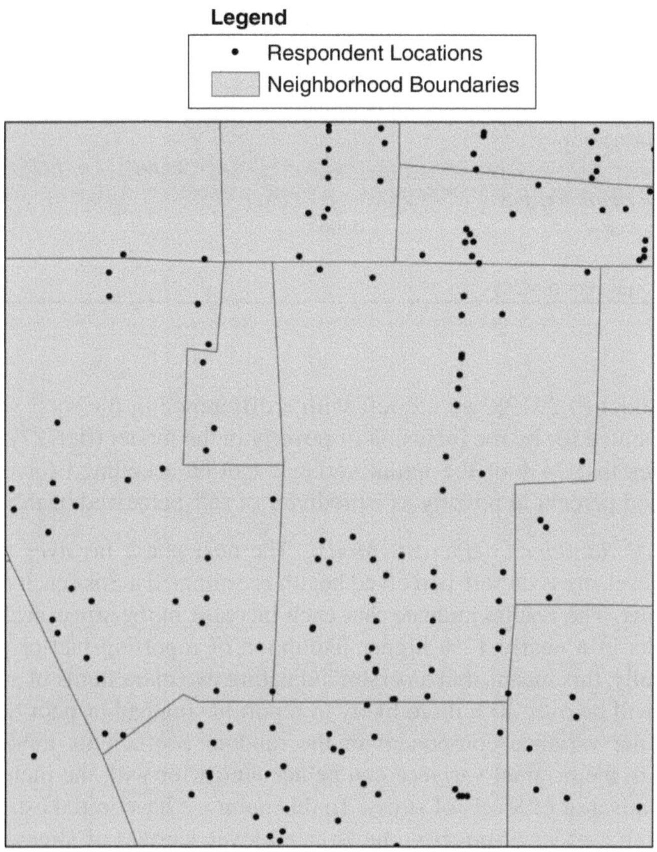

Fig. 8.1 Example result of geocoding survey respondents to their actual physical address within neighborhood boundaries; Panel of American Life Survey (PALS)

outcome would neglect to take into account a significant amount of variation that can be accounted for by neighborhood factors. Such environmental effects have become commonplace in epidemiology and continue to gain in presence in many other sociologically oriented fields.

Step 2: Regression with Means as Outcomes Model. In the second step of this HLM analysis, we examine the mean effects of neighborhood poverty levels on the self-perceived health. The results are presented in Table 8.1. For each percent increase in the percent in poverty in the neighborhood, the likelihood of perceiving one's health to be poor or bad increases by about 3%. Thus, a hypothetical situation in which a neighborhood has a higher level of poverty by 10% of the population than another neighborhood would result in a 30% higher likelihood of those residents reporting bad or poor health. Turning to the next statistic, the variance component of the Means as Outcomes Model is 0.2192. If we use the equation presented in Chap. 7 and subtract that from the variance component obtained in the one-way

Table 8.1 Odds ratios of aspirations of upward social mobility by class background within and HLM framework

	Means as outcomes	Random coefficients	Intercepts as slopes
Individual characteristics			
Stress level	–	1.1667*	1.2612*
Neighborhood characteristics			
Percent in poverty	1.0306*	–	1.0437*
Cross-level interactions			
Stress level × percent poverty	–	–	1.0091*

*$p < 0.001$

ANOVA model (0.2919), we are left with a difference in the total variance that can be accounted for by the inclusion of poverty in the model (0.0727). Ultimately, this indicates that 24% of the initial variance can be accounted for by including neighborhood percent in poverty as a predictor of self-perceived health.

Step 3: HLM Random Coefficients Model. The next phase involves the effect of individual level stress on self-perceived health is estimated across each of the neighborhood units. The results indicate that each increase in the stress scale (ten point scale) results in a nearly 17% higher likelihood of reporting bad or poor health. Hypothetically, this means that a person indicating two extra items of stress in their daily lives will be over 30% more likely to report having bad or poor health. When examining the variance component in the random coefficients model (0.1951), about 33% in the original variance can be accounted for with the inclusion of this individual indicator of levels of stress. To this point we have initial estimates of the isolated level 2 effect of poverty, the isolated level 1 effect of stress, and a good idea of the proportional impact each accounts for in the variation of respondents self-perceptions of their own health.

Step 4: HLM Intercepts and Slopes-as-Outcomes Model. The final step within this HLM framework is to test for potential cross-level interactions between the individual indicator of stress and the ecological level of poverty in the resident's neighborhood. The results in Table 8.2 report the updated main effects of each and an interaction estimate. Specifically, the effects of individual level stress now increase the likelihood of reporting poor or bad health by almost 26% for each increase in item answered as causing stress in their lives. Moreover, the likelihood of reporting poor or bad health increases by about 4% for each increase in the percent of their neighborhood that lives in poverty. These results are actually stronger than the isolated results of the *Means as Outcomes* and *Random Coefficients* models. However, of most interest here is the cross-level interaction between the two variables. Here we see that for each increasing percentage of the neighborhood in poverty, the effect of additional stress items further increases the likelihood of reporting poor or bad health by about 1 one additional percent. Thus, the effect of individual level stress and neighborhood poverty are independently significant predictors of self-perceived health, but also the interaction of individual and ecological effects proves to further predict variation in individual level health perceptions.

Table 8.2 Within and between neighborhood variance component indicators

	St. dev.	Variance component	df	Chi-square	P-value
Between neighborhoods	0.5402	0.2919	297	426.75	<0.001
Within neighborhoods (individual)	0.5432	0.1951	297		
Total		0.4870			

These results provide support for the idea that the ecological effect has its own unique effect on individual social and behavioral outcomes, but also has an interacting effect with individual level personal and demographic characteristics. Thus, ignoring such relationships neglects to understand such main effects and their association to individual level relationships. The next section will present another example analysis of point pattern data pertaining to crime data in the city of Atlanta, GA.

8.3 Detecting Spatial Clusters of Events: Atlanta, GA Homicide Locations, 2004–2009

Here we provide a short example analysis of the geographic location of events in the form of a point pattern analysis of homicides in Atlanta from 2004 to 2009.[1] The analysis is specifically focused on issues of proximity. In this respect, we are interested in the proximity of these *event locations* in reference to one another. As indicated in Chap. 7, one of the primary methods of analyzing point patterns is to test for deviations from a uniform distribution to a clustered pattern. Using the software *Crimestat,* we have identified a series of clusters in the point pattern presented in Fig. 8.2. The figure is organized so that all homicides in the time period are presented in relation to underlying geography compatible with the appending of census data from the "block-group" (see U.S. Bureau of the Census GARM 1994). Block-groups are smaller geographies than census tracts, but one can see that if data on the social characteristics of these areas was added, we could make inductive findings that might lead to more hypotheses and more sophisticated predictive modeling procedures. Here we only focus on the identification of deviations from a completely random pattern of points. Sometimes a point-map can seem obvious as to the pattern but human eyes can be deceived into "seeing" what is expected. That is one key element of exploratory spatial data analysis (ESDA) to "put numbers back on the map" to caution against more subjective interpretations.

In addition to the events and underlying geography, the figure also includes descriptive indicators of the spatial mean and significant clusters of high levels of homicide. In regards to the spatial mean, one can see that the average place where crime occurs is actually in downtown Atlanta. The spatial mean is a good tool to allow researchers to see if the data is heavily concentrated in specific spatial areas

[1] These data were obtained by special request from the GIS Unit of the Atlanta Police Department.

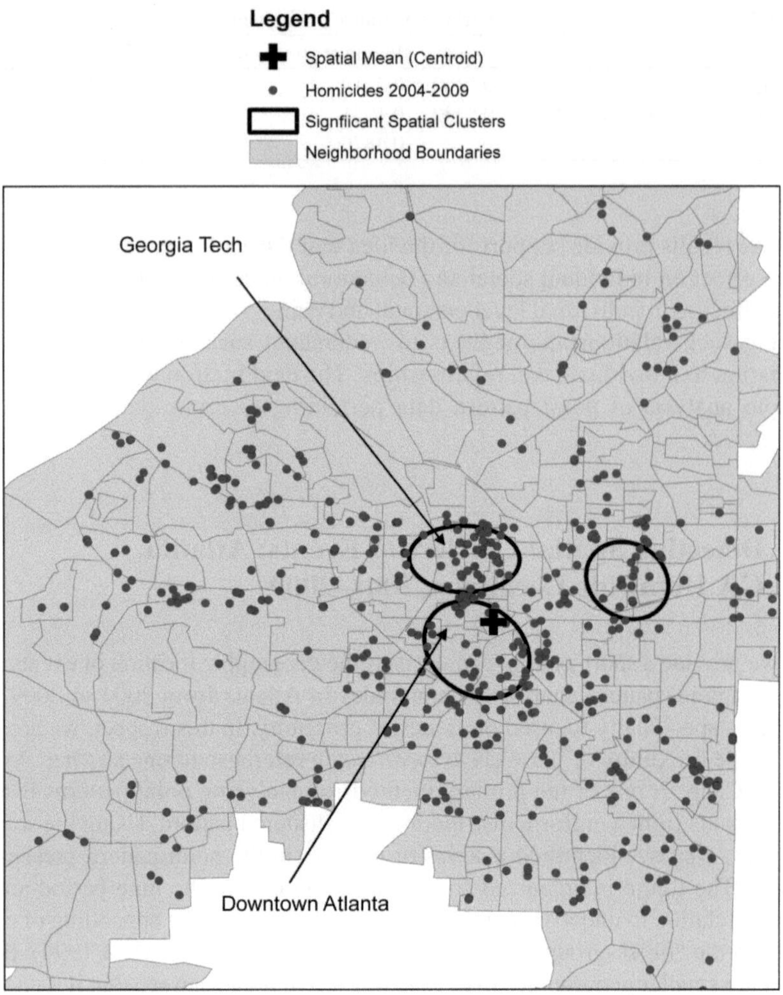

Fig. 8.2 Spatial point pattern analysis, Atlanta, GA area homicides 2004–2009

as those areas will influence the location of the spatial mean by giving heavier weight to the those sides with a higher density of events. Thus, it should be interpreted in a similar fashion as a statistical mean in that it demarcates the average location of crime.

Similar to the average, the ellipses, which represent significantly high levels of homicide during the time period, are also based on densities and proximities in relation to other events on the spatial plane. In the case of the Atlanta area, three significant clusters emerge. First, a cluster has been identified just north of downtown and near Georgia Tech. Just south of that area, and encompassing the spatial mean as well as a core part of the downtown business district in Atlanta, is the second significant cluster of homicides. Finally, a third area of significantly high homicide incidence appears to the east of the other two clusters. These three clusters represent statistically high levels of homicides, both in number and density, compared to what would be expected under a random distribution of points. Of course, they are also all located very near the most densely populated areas of the city during the day time which might account for the higher numbers (Conversion into homicide rates could be conducted) However, given the identification of spatial clustering, further analyses could potentially build on this to incorporate predictive models of future events and the underlying socio-cultural characteristics of the neighborhoods in which these events occur (see Osgood 2000).

8.4 Modeling Contiguous Regions of Well-Being: A Spatial Regression Approach

Our final example makes use of spatial clustering and regression techniques taking into account contiguous adjacency as a measure of spatial closeness. Here we present data collected from two public sources at the U.S. county level. Our example analysis takes aim at testing one of the most fundamental social relationships in the literature, the protection effect of being married. In our analysis, however, we extend this hypothesis to the aggregate level with a research hypothesis aimed at investigating the relationship between community level marriage rates and related levels of community health. The health indicator will be measured via the average life expectancy of the population within the county. It is well documented that regions of the country vary significantly in regards to the health and well-being of the populations that inhabit those regions. Here we aim to investigate the degree to which observed variations in the marriage rate of communities contributes to these spatial variations.

Step 1: Detecting Spatial Dependence and Local Levels of Association. The first step in determining the appropriate approach to testing this relationship is to test for the potential existence of spatial clustering among U.S. counties in regards to their levels of life expectancy via Exploratory Spatial Data Analysis (ESDA) techniques. As discussed in Chap. 7, we employ the global Moran's I statistic as our indicator of spatial clustering. The results of the analysis are presented in Fig. 8.3. The statistic in the figure indicates that a strong level of positive spatial clustering exists, per the 0.717. This can be interpreted similar to a correlation coefficient and the statistic represents the degree to which each counties level of life-expectancy is associated with their neighbor's life expectancy. Thus, a strong positive statistic, such as the one here, indicates that counties are very similar to their neighboring counties in regards to life expectancies. It is important here to note that a counties neighbor is any county that shares a contiguous border with that county. Thus, any counties that are contiguous in any direction are considered neighbors. The analysis, which was undertaken in the GeoDa, allows for the creation of theoretically appropriate spatial weights. In different cases, a distance based, number of nearest neighbors based and even directional contiguity based weight matrix may be more appropriate.

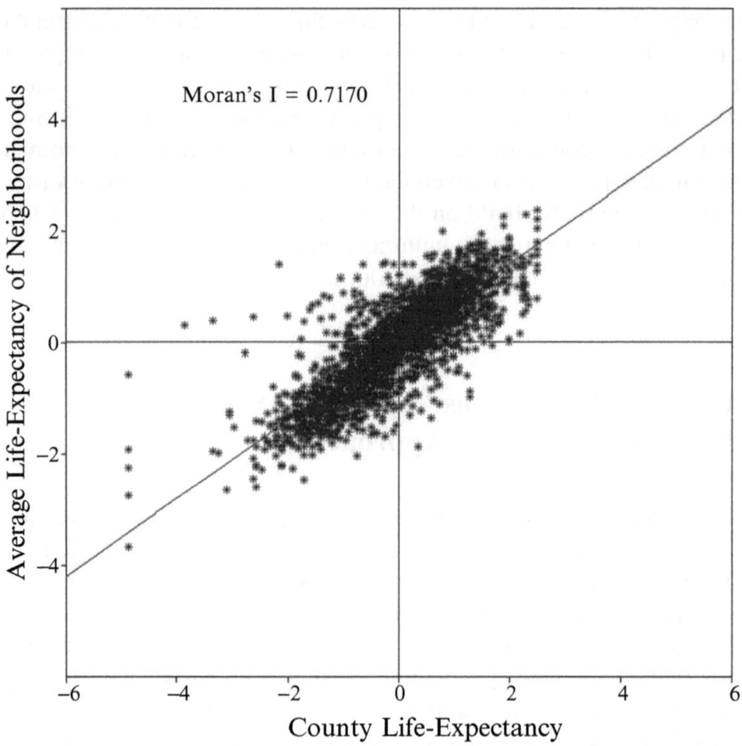

Fig. 8.3 Global Moran's I scatterplot detecting spatial clustering of life-expectancy values across U.S. counties

Given the results of the ESDA, we map the levels of life expectancy and further identify spatial clusters in Fig. 8.4. In the top panel of the figure, one can easily see that the lowest levels of life expectancy are disproportionately in the American South. In contrast, the Northeast coast and the upper Midwest have visibly higher proportions of counties in the highest level of life expectancy. While these patterns are visibly apparent, in order to test for the actual locations of significant clusters of high and low life expectancy we also map the results of Anselin's Local Indicator of Spatial Association (LISA) test (Anselin 1995). The LISA statistics identifies statistically significant areas of high levels of life expectancy and low levels of life expectancy by testing the degree to which each county is similar to its neighbors. The LISA statistic also provides information on counties that are high or low in levels of life expectancy but surrounded by neighboring counties that are the opposite. This would indicate negative levels of spatial autocorrelation. Given the strong and positive levels of autocorrelation, per the Moran's I statistic in Fig. 8.3, we would expect to have very few of those clusters. As a result the bottom panel presents a spatial illustration of local clusters of high and low levels of life expectancy. The significant clusters of high life expectancy have diagonal lines running from lower

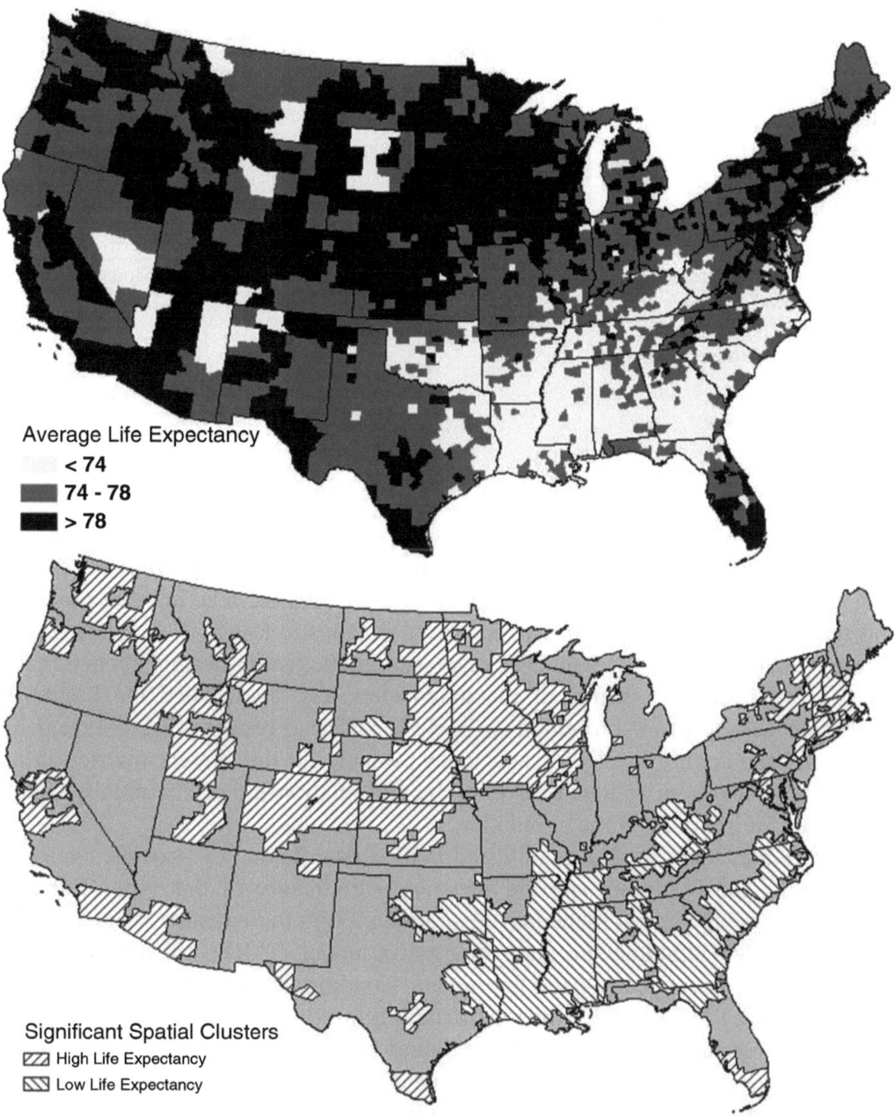

Fig. 8.4 Spatial variation of average life expectancy and local significant clusters

left to upper right, while the significant clusters of low life expectancy have diagonal lines running from upper left to lower right. The results pretty much mirror what we might expect from the visual inspection of the upper panel.

Step 2: Identify Appropriate Regression Control and Weight Subsequent Regression Analyses. Once the determination is made that significant levels of spatial dependence exist, it becomes important to control for the effect of adjacency in order to

Table 8.3 Diagnostics for spatial dependence

Test	MI/DF	Value	Probability
Moran's I (error)	0.695	N/A	N/A
Lagrange Multiplier (lag)	1	722.23	0.000
Robust LM (lag)	1	3.817	0.057
Lagrange Multiplier (error)	1	4179.74	0.000
Robust LM (error)	1	3461.33	0.000

Table 8.4 Unstandardized bivariate spatial regression results predicting average life expectancy by percent married

	Model estimates
Percent married	0.101*
Spatial weight (ϵ)	0.869*
R-square	0.686
AIC	10142.3

ϵ spatial error model (Simultaneous Auto-regressive Model)
*P-value < 0.001

ensure that traditional assumptions of linear regression are not violated. As discussed in Chap. 7, there are two types of spatial autoregressive models. In order to detect the type of underlying spatial dependence, a diagnostic test must be under taken through the running of a classic OLS regression model with a weight opened in the GeoDa program. The results of the diagnostics test, also called the Lagrange Multiplier LM test, are presented in Table 8.3. These results were obtained by regressing the county-level life expectancy variable on the percent married in an attempt to uncover an aggregate level protection relationship of community health given rates of marriage in the population.

In the table, a series of tests are listed in the first column, an associated value is listed in the third column, and the statistical significance of that type of spatial dependence being present in the far right column. From these results, it appears that the error model (Simultaneous Auto-Regressive model: SAR) is the most efficient, although there is support for the lag model (Conditional Auto-Regressive model: CAR). The way that is determined is by identifying the LM test that reports being significant. Again, in this case, there is support for both, however, the results of the Robust LM (lag) test are borderline significant and thus provide enough of a reason to incorporate the SAR modeling procedures in subsequent spatially-weighted regression models.

The results from the simple bivariate SAR model are presented in Table 8.4. From the results, it is evident that there may in fact be an aggregate level protection relationship between the two variables. The unstandardized coefficient estimate of 0.101 indicates that for every percent increase in the marriage rate, the average life expectancy of a county increases by about a tenth of a year (or a little over a month). Thus a person from a county with a 12% higher marriage rate than another county would on average enjoy an extra year of life. Of course, this is not a traditional regression model in that it does not include any controls so much of this relationship is likely related to other factors that have been shown to be associated with high marriage rates (i.e. high family incomes, lower percent minority, better education levels, etc.). An additional interest here is the effect of the spatial weight parameter (ε) is 0.869. This indicates that the effect of being a neighboring county is stronger than the percent married in the county in regards to its prediction of life expectancy levels. Furthermore, by simply taking into account the fact that places closer together are more alike than places far apart, we have explained nearly 70% of the variation in county level life expectancy rates with only adjacency and the percent married ($R^2=0.686$).

8.5 Conclusion

The example analyses presented in this chapter were intended to present popular methods for the examination of spatial concepts like containment, proximity, and adjacency. As mentioned in the previous chapter, many other methods exists for taking these relationships into account and both authors strongly encourage further reading on the topics to gain a more complete understanding of what is available. The current scope of this book is limited in that sense, however, the methods of geo-sociology continue to advance and become more and more mainstream. Here we hope only encourage further reading and an interest in the usage of spatially-centered methodologies by providing generally interesting examples of poverty, crime, health and well-being and the methods that may be used to examine the degree to which the ecological environment and the social phenomena interact. Finally, this chapter represents the final chapter of Part II of this book. The following chapter will conclude the text with a focus on discussion of the book in its entirety and a brief comment on future directions in the geo-sociological theory and methods.

Chapter 9
Concluding Remarks

As the preceding chapters have documented, the history of ecological analysis in the field of sociology is one that goes back over a century. It is arguably also the core foundation which from which American Sociology has developed. Much of this development can be linked back to the Chicago School of urban and community sociology. We document through bibliometric study that the Chicago School heritage should be formally acknowledged as building upon the work of classical location theories developed those in the field of economic and from the work by rural sociologist Charles J. Galpin at the University of Wisconsin-Madison (Park 1929a, b). This legacy was forgotten, however, by the time that the seminal work, *The City* (1925) was published and still not acknowledged a century later (Abbott 1999). We hope that this small volume helps in some way to facilitate a clearer understanding of that heritage in hopes of continuing to spur future developments in the area.

Other important developments include early economic theories of location and the introduction of the isotropic map as a tool for social research, first by Galpin (1915) and later by other rural sociologists through its application to understanding core social problems. Since those early innovations, theories of location and spatially-centered analytic techniques developed in substantial ways, especially during the past two decades. Of importance to this book is the fact that the development of these theories and methods did not occur in historical unison. Instead, the theoretical foundations, which still today provide much of our understanding of macrosociological group processes, were developed decades before the computational and analytical capabilities to test such tenants were possible. We believe that they are now not only possible but commonly available to virtually all sociologists.

Ironically, given this unmatched historical development, the current boom in spatially-centered analyses finds itself unprepared to explain theoretically relationships associated with spatial concepts. For example, we have dedicated some narrative explanation of the core spatial concepts of *containment*, *proximity*, and *adjacency*. Today there are methodological tools to operationalize and test the effects of being contained within a specific unit (i.e. neighborhood, county, etc.) and being spatially proximate or adjacent to certain social stimuli. Here we have

J.R. Porter and F.M. Howell, *Geographical Sociology: Theoretical Foundations and Methodological Applications in the Sociology of Location*, GeoJournal Library 105, DOI 10.1007/978-94-007-3849-2_9, © Springer Science+Business Media B.V. 2012

attempted to bring together some of the most important early developments in locational theory with contemporary methodological developments in spatially-centered techniques, which in combination contribute to *geo-sociology* as a substantive framework. As this is a first attempt at such a consolidation, we expect to be met with certain agreement and opposition. However, in both cases our hope is to spur a discussion which may further lead to a better understanding of the effects of context, space, place, and geography in the everyday interactions of individuals and the social processes that occur at their aggregated levels (see also Lieberson and Lynn 2002).

What should a spatial analysis in sociology contain? To begin, we argue that a spatial analysis should be:

Spatial in the middle-range *theoretical framework*;
Spatial in one or more key *concepts*;
Spatial in the *operationalization* of concepts; and
Spatial in the *analytical procedures* used to explore or test the framework.

As Howell (2004) noted, this is a high standard to reach in practice but one to elicit a few "good theories of the middle-range" (Merton 1968). Indeed, our own published work currently fails in many ways to fully reach each element of this standard. Our exhortation is for sociologists to both acknowledge this spatial heritage, learn the methods required to embrace it, and move the sub-field of ecological sociology forward in substantive and theoretical terms.

This book began with a definition of the term *geo-sociology* in both practice and principle with a special focus on its situation within the larger field of sociology. Following that introduction, Part I of this book described the historical development of location theory from the early focus on communities in isolation, by economists, to the more modern focus on the relationships between man and the environment in an increasingly global economy. The second part of this book (Part II) built on these theoretical understandings of the relationships between humans and their geographic locations by focusing contemporary methodological approaches that allow sociologists to measure and test the spatial concepts that currently remain unexplored and abstract in the field of sociology.

This is where we feel that the most significant gains can be made in the field of geo-sociology in the near term. To date, methods for understanding the relationship between humans and their geographic context continue to be developed by researchers at institutes and centers such as the GeoDa Center for Geospatial Analysis and Computation at Arizona State, the Center for Spatially Integrated Social Science (CSISS) at the University of California Santa Barbara, the GeoVista Center at Pennsylvania State University, and in various university centers in the United Kingdom (e.g., National Centre for Geocomputation at the National University of Ireland-Maynooth, Centre for Advanced Spatial Analysis at the University College London), and Japan (Center for Spatial Information Science at the University of Tokyo). As these methodological tools allow us to analytically link individuals and groups to geography in a way that was previously not possible, we find ourselves faced with more questions concerning the theoretical relationships that we continue to uncover.

What is needed is a focus on *linking* the current movement of contextual sociology (research practice) to a well-defined rhetoric for the examination of social problems that have long been of interest to sociologists. To that point, the rhetoric regarding the importance of context is no different than those that founded the first departments of sociology, nor the general definitions of sociology that we still read to today in our most popular introductory textbooks. What *is* different is that leading founders of the discipline had a *clear intellectual connection* between their rhetoric and their research practice. Today we are faced with more computational sophistication than ever before, and trend is only moving forward in this area. As one can surely tell from our enthusiasm in the undertaking of this book project, we believe that incorporating geographic variations in variants of the social and physical environment is essential to much of the research we undertake as sociologists. However, our current ability to put people into these ecological settings, through the implementation of many of the methods described in this book (and many others that were not) does not necessarily mean that the reasons for doing so are well understood.

As Voss (2007) soberly noted regarding contemporary practice, much has been said but little has yet to be done. Following in the footsteps of Voss (and others) we continue to call for a greater acceptance of spatial analyses in sociology and a further development of the theoretical frameworks that will allow us to understand such relationships. Spatial analysis in mainstream sociology is about where path analysis was in the early 1970s: rapidly evolving methods in search of more substance. Such trends signal growing opportunities for those interested in pushing forward geo-social approaches to research in sociology in much the same way that similar opportunities existed, and were met, in the incorporation of other out-of-discipline methods that have been successfully incorporated into mainstream sociology.

In general, this need has been recognized and calls for attention to its development have been made. Such calls have been made since the late 1990s when the National Research Council report, *People and Pixels*, signaled a need to respond as a discipline. As noted above, the NSF-sponsored Center for Spatially Integrated Social Science (www.csiss.org) effort is a major response, which has provided a wealth of information to spur spatial thinking in the social sciences at-large. Goodchild et al. (2000) argued that the spatial domain puts the social sciences on a common lexicon to communicate. In the case of sociology, the ability to communicate across disciplines is especially valuable given its long history of incorporating theory and methods from other social science concentrations (e.g., path analysis by Otis Dudley Duncan).

In order to flourish, geo-sociology is in need of a few good theorists of the middle-range (Howell 2004). One example of such theory is the work by Sampson and colleagues (1997) which linked levels of neighbor efficacy to variations in crime. This type of an understanding helps researchers to continue to develop theories of concepts like containment. For instance, what is it about being in a neighborhood of high collective efficacy that limits criminal behavior? Sampson and colleagues have provided strong clues as to what that relationship is and others continue to push this understanding forward.

Another necessity in the continued development of geo-sociology includes methodological advances involving the growth of spatially-referenced data, methods, and computing capacity to conduct analyses. In the past few years, this need has begun to be addressed as many large scale data collection efforts now include geographic identifiers of both the respondent's locations and the units in which the respondent is contained for the purpose of appending administrative data. Such efforts provide some of the data used in the example applications in Chaps. 6 and 8. For instance, some of the data used in the Chap. 6 example analysis is from the Louisville Metro Survey, which is collected on an annual basis by the University of Louisville Sociology Department and provides information on the nearest cross-streets of its respondents in selected years. In addition, the Panel of American Life Survey (PALS) data (used in the HLM example in Chap. 8) is collected with restricted geo-ids that allow for the linkage of the survey respondents to their county, neighborhood, and actual block through the geocoding of the respondents obtained physical street location. In both examples, the ability to link the survey respondents to a larger geographic context has provided researchers with a greater ability to understand social relationships beyond the individual level data that is provided in the survey questionnaire.

The newness of geo-social approaches to research in sociology has also been met with some hesitancy in acceptance, which is another obstacle that must by overcome for continued growth. This issue continues to become less and less of an issue as the methodological tools become more popular and commonplace in the research journals that sociologist regularly read.[1] The taking of theoretical and methodological "chances" predicated upon creative spatial thinking and the acceptance of such approaches by gatekeepers to the published literature (editors, reviewers) will continue to ameliorate this issue as spatial thinking becomes more commonplace. Finally, journals must also be willing to complement such an approach through a transitioning to publish color graphics since it remains difficult to convey enough information in current grayscale format.

One way to ensure that such approaches continue to grow in the field is through the implementation of training in spatial methods at both undergraduate and graduate levels. Over the last decade the availability of such courses has gone from almost non-existent in sociology departments to be much more commonplace. In fact, many Ph.D.-granting departments in sociology now offer geo-sociologically relevant advanced courses in some form. This advanced training at the graduate level especially makes appropriate spatial methods as much "ours" as any other discipline. An historical metaphor is that the Goodness of Fit Chi-squared statistic was not "sociology" when Karl Pearson first published it for use in the statistics literature. It is now taught in undergraduate statistics courses in the sociology major. Spatial analysis in the published literature is likely to continue to push the current boundaries

[1] However, just recently one of the authors received a negative decision on a submitted manuscript with the primary reason being that "the methods employed in the analysis are not for a general sociology audience".

of what is "considered sociology" and therefore integrated into the curricula of university departments of sociology.

We have argued elsewhere (Howell 2004) that there are some critical elements of curriculum content for adequate spatial analysis coursework. These include: data visualization (GIS), Census geography, spatial data (TIGER database and related), and spatial statistics. These elements are reflexive in nature, requiring that they be learned as a set. For instance, many GIS-based visualizations are based upon poorly-understood "geographies" which misappropriate available spatial data. Spatial statistical analyses without some visualization of the data can result in inappropriate interpretations of the analysis. The lack of an adequate understanding of key digital databases, especially the Census Bureau's TIGER database, can result in either missed research opportunities or inappropriate understandings of the data used in either data visualization (GIS), statistical analysis (spatial statistics), or even the associated attribute data themselves (e.g., errors in merging data to the appropriate TIGER or other polygon, line, or point data). Our own experiences in teaching this material since 1997 leads us to strongly encourage readers to consider these domains of instruction in building new course, whether they are taught in one course or in a sequence of courses.

Overall, the development of a geo-sociology has been predicated on the foundations of sociology itself. After all, sociology is the study of society and its effect on individuals and groups within that particular social setting. Inherent in this foundation is the potential variations in human behaviors and group processes that occur *across space and over time*. A geographic focus in sociology is important and can be traced to the roots of the discipline itself. As geo-sociologists, we look forward to the continued development of the approach in the coming years. In particular, we hope that this book follows Leiberson and Lynn's (2002) strategy and helps to spur discourse in the substantive area, one that we hope to be modest contributors.

Bibliography

Abbott, A. (1994). History and sociology: The lost synthesis. In E. H. Monkkonen (Ed.), *Engaging the past: The uses of history across the social sciences*. Durham/London: Duke University Press.
Abbott, A. (2001). *Time Matters: On Theory and Method*. Chicago: University of Chicago Press.
Abbott, A. D. (1999). *Department and discipline: Chicago sociology at one hundred*. Chicago: University of Chicago Press.
Adams, C. C. (1935). The relation of general ecology to human ecology. *Ecology, 16*, 316–335.
Albrecht, D. E., & Murdock, S. H. (1990). *The sociology of agriculture: An ecological perspective*. Ames: Iowa State University Press.
Alonso, W. (1964). *Location and land use*. Cambridge: Harvard University Press.
Anderson, M. S. (1951). *Geography of living things*. Liverpool: C. Tinling and Co.
Anselin, L. (1988). *Spatial Econometrics: Methods and Models*. Dordrecht: Kluwer Academic Publishers.
Anselin, L. (1990). Spatial dependence and spatial structural instability in applied regression analysis. *Journal of Regional Science, 30*(2), 185–197.
Anselin, L. (1995). Local indicators of spatial association – LISA. *Geographical Analysis, 27*, 93–115.
Anselin, L., & Bera, A. (1998). Spatial dependence in linear regression models with an introduction to spatial econometrics. In A. Ullah & D. E. Giles (Eds.), *Handbook of applied economic statistics* (pp. 237–289). New York: Marcel Dekker.
Anselin, L., & Tam Cho, W. K. (2002a). Spatial effects and ecological inference. *Political Analysis, 10*, 276–297.
Anselin, L., & Tam Cho, W. K. (2002b). Conceptualizing space: Reply. *Political Analysis, 10*, 301–303.
Barber, C. B., Dobkin, D. P., & Huhdanpaa, H. T. (1996). The Quickhull algorithm for convex hulls. *ACM Transactions on Mathematical Software, 22*(4), 469–483.
Bickel, R. (2007). *Multilevel analysis for applied research: It's just regression*. New York/London: The Guilford Press.
Blanchard, T. C., & Matthews, T. L. (2007). Retail concentration, food deserts, and food-disadvantaged communities in rural America. In C. C. Hinrichs & T. A. Lyson (Eds.), *Remaking the North American food system: Strategies for sustainability*. Lincoln: University of Nebraska Press.
Blau, P. M., & Duncan, O. D. (1967). *The American occupational structure*. New York: Wiley.
Bogue, D. J. (2009, November). A tribute to Amos Hawley: Mentor, colleague and friend. *Footnotes of the American Sociological Association, 37*(8). http://www.asanet.org/footnotes/nov09/tribute_1109.html

J.R. Porter and F.M. Howell, *Geographical Sociology: Theoretical Foundations and Methodological Applications in the Sociology of Location*, GeoJournal Library 105, DOI 10.1007/978-94-007-3849-2,

Booth, C. (1887). The Inhabitants of the Tower Hamlets (School Board Division), Their Condition and Occupations. *Journal of the Royal Statistical Society 50*, 326–401.

Booth, C. (1889–1903). *Labour and the life of the people* (Vols. 17). London: Macmillan and Co.

Boucek, B., & Moran, E. F. (2004). Inferring the behavior of households from remotely sensed changes in land cover: Current methods and future directions. In: M. F. Goodchild and D. G. Janelle (Eds.), *Spatially Integrated Social Science 23–47.* New York: Oxford University Press.

Bourdieu, P. (1983). Men and machines. In K. Knorr-Cenina & A. V. Cicourel (Eds.), *Advances in social theory and methodology*. Boston: Routledge & Kegan Paul.

Brown, L. D., Gans, N., Mandelbaum, A., Sakov, A., Shen, H., Zeltyn, S., & Zhao, L. H. (2005). Statistical analysis of a telephone call center: a queueing science perspective, *Jour Amer Statist Assoc 100*, 36–50.

Burgess, E. W. (1930). The value of sociological community studies for the work of social agencies. *Social Forces, 8*(4), 481–491.

Butell, F. H. (1976). Social science and the environment: Competing theories. *Social Science Quarterly, 57*, 307–323.

Buttell, F. H. (1986). Sociology and the environment: The winding road toward human ecology. *International Social Science Journal, 38*, 337–356.

Butell, F. H., Dickens, P., Dunlap, R. E., & Gijswijt, A. (2002). Sociological theory and the environment: An overview and introduction. In R. E. Dunlap, F. H. Buttel, P. Dickens, & A. Gijswijt (Eds.), *Sociological theory and the environment: Classical foundations and contemporary insights*. Lanham: Roman and Littlefield.

Campbell, A. J., & Vogt, P. L. (1918). *Introduction to Rural Sociology*. New York: Appleton Co.

Catton, W. R. (1994). Foundations of human ecology. *Sociological Perspectives, 37*, 75–95.

Catton, W. R. (2002). Has the Durkheim legacy misled sociology? In R. E. Dunlap, F. H. Buttel, P. Dickens, & A. Gijswijt (Eds.), *Sociological theory and the environment: Classical foundations and contemporary insights*. Lanham: Roman and Littlefield.

Catton, W. R., & Dunlap, R. E. (1978). Environmental sociology: A new paradigm. *The American Sociologist, 13*, 41–49.

Chang, K. (2006). *Introduction to geographic information systems* (3rd ed.). Boston: McGraw-Higher Education.

Christaller, W. (1933). *Central places in Southern Germany* (C. E. Baskin, 1966, Trans.). Engle Cliffs: Prentice-Hall.

Cohen, J., & Tita, G. (1999). Diffusion in homicide: Exploring a general method for detecting spatial diffusion processes. *Journal of Quantitative Criminology, 15*(4), 451–493.

Coleman, J. S. (1994). Social Capital in the Creation of Human Capital. *The American journal of sociology.*

Collins, R. (1983) In K. Knorr-Cenina & A. V. Cicourel (Eds.), An empirical relativist programme in the sociology of scientific knowledge. *Advances in social theory and methodology.* Boston: Routledge & Kegan Paul.

Craglia, M., Haining, R., & Signoretta, P. (2001). Modeling high-intensity crime areas in English cities. *Urban Studies, 38*(11), 1921–1941.

Cressie, N. A. (1993). *Statistics for spatial data.* New York: John Wiley & Sons.

de Smith, M. J., Goodchild, M. F., & Longley, P. A. (2007). *Geospatial analysis: A comprehensive guide to principles, techniques, and software tools* (2nd ed.). Leicester: Matador Press.

Dear, M. (1988). The postmodern challenge: Reconstructing human geography. *Transactions of the Institute of British Geographers, 13*(3), 262–274.

Duncan, O. D. (1966). Path analysis: Sociological examples. *The American Journal of Sociology, 72*(1), 1–16.

Duncan, O. D., & Schnore, L. (1959). Cultural, behavioral, and ecological perspectives in the study of social organization. *The American Journal of Sociology, 65*, 132–146.

Dunlap, R., & Catton, W. (1979a). Environmental Sociology. *Annual Review of Sociology 5*, 243–273.

Dunlap, R. E. (2002). Paradigms, theories, and environmental sociology. In R. E. Dunlap, F. H. Buttel, P. Dickens, & A. Gijswijt (Eds.), *Sociological theory and the environment: Classical foundations and contemporary insights*. Lanham: Roman and Littlefield.

Durkheim, E. (1897). Note sur la Morphologie Sociale. *L'Anne'e Sociologique, 2*, 521–522.

Durkheim, E. (1909). Sociologie et Sciences Sociales. In J.-C. Filloux (Ed.), *La Science Sociale et l'Action*. Paris: Presses Universitaires de France.

Duster, T. (1983). Intermediate steps between micro- and macro-integration: The case of screening for inhereted disorders. In K. Knorr-Cenina & A. V. Cicourel (Eds.), *Advances in social theory and methodology*. Boston: Routledge & Kegan Paul.

Elliot, P. E., Cuzick, J., English, D., & Stern, R. (1996). *Geographical & environmental epidemiology: Methods for small-area studies*. Oxford: Oxford University Press.

Ellwood, C. A. (1927). Recent developments in sociology. In E. C. Hayes (Ed.), *Recent developments in the social sciences*. Philadelphia/London: J.B. Lippincott Company.

Entwisle, B. (2007). Putting people into place. *Demography, 44*(4), 687–703.

Entwisle, B., Walsh, S. J., Rindfuss, R. R., & Chamratrithirong, A. (1998). Land-use/land-cover and population dynamics, Nang Rong, Thailand: Chapter 6. In *People and pixels*. Washington, DC: National Academy Press.

Foody, G. M. (2006). GIS: Health applications. *Progress in Physical Geography, 30*(5), 691–695.

Fotheringham, A. S., Brundson, C., & Charlton, M. (2002). *Geographically weighted regression: The analysis of spatially varying relationships*. West Sussex: Wiley.

Frey, F. C., & Smith, L. T. (1936). The Influence of the AAA Cotton Program Upon the Tenant, Cropper, and Laborer. *Rural Sociology, 1*(4), 493.

Friedman, S. W. (1996). *Marc bloch, sociology, and geography*. Cambridge: Cambridge University Press.

Friedman, T. L. (2006). *The world is flat: A brief history of the twenty-first century, updated and expanded*. New York: Farrar, Straus and Giroux.

Frisbie, P. (2009). A tribute to Amos Hawley: Mentor, colleague and friend. *Footnotes of the American Sociological Association, 37*(7).

Fuller, S. (2006). *The new sociological imagination*. London: Sage.

Galpin, C. J. (1915). *The social anatomy of an agricultural community*. Research Bulletin Num. 34. Madison: University of Wisconsin Agricultural Experiment Station.

Galpin, C. J. (1918). Rural sociology: Standardization of research co-operative plan of national rural research. *The American Journal of Sociology, 24*(3), 303–310.

Galpin, C. J. (1930). Rural life. *The American Journal of Sociology, 35*(6), 1010–1016.

GARM, (1994). *Geographic Area Reference Manual*. U.S. Census Bureau: Washington, DC.

Giddens, A. (1983). Agency, institution, and time-space analysis. In K. Knorr-Cenina & A. V. Cicourel (Eds.), *Advances in social theory and methodology*. Boston: Routledge & Kegan Paul.

Goodchild, M. F., & Janelle, D. G. (2004a). Thinking spatially in the social sciences: Chapter 1. In M. F. Goodchild & D. G. Janelle (Eds.), *Spatially integrated social sciences*. Oxford: Oxford University Press.

Goodchild, M. F., & Janelle, D. G. (Eds.). (2004b). *Spatially integrated social science*. New York: Oxford University.

Goodchild, M. F., Anselin, L., Appelbaum, R. P., & Harthorn, B. H. (2000). Toward spatially integrated social science. *International Regional Science Review, 23*(2), 139–159.

Goodman, L. A. (1953). Ecological regressions and behavior of individuals. *American Sociological Review, 18*, 663–664.

Goodman, L. A. (1959). Some alternatives to ecological correlation. *The American Journal of Sociology, 64*, 610–625.

Grannis, R. (1998). The importance of trivial streets: Residential streets and residential segregation. *The American Journal of Sociology, 103*(6), 1530–1564.

Grannis, R. (2005). T-communities: Pedestrian street networks and residential segregation in Chicago, Los Angeles, and New York. *City and Community, 4*(3), 295–321.

Greve, H. R. (2002). An ecological theory of spatial evolution: Local density dependence in Tokyo banking, 1894–1936. *Social Forces, 80*(3), 847–879.

Gross, M. (2004). Human geography and ecological sociology: The unfolding of human ecology. *Social Science History, 24*(4), 575–605.

Haeckel, E. (1866). *Generelle Morphologie der Organismen*. Berlin: Reimer.

Hagerstrand, T. (2000). The computer and the geographer. *Transactions of the Institute of British Geographers, 42*, 1–19.

Haggert, P. (1967). Three pioneers in spatial theory. *The Geographical Journal, 133*(3), 357–359.

Harter, R. (2010). *Design and Operational Changes for REACH US*. Presented at the American Statistical Association Annual Meetings: Survey Research Methods Section.

Hawley, A. (1950). *Human ecology: A theory of community structure*. New York: Ronald Press.

Hawley, A. (1986). *Human ecology: A theoretical essay*. Chicago: University of Chicago Press.

Hayes, E. C. (1908). Sociology and psychology; sociology and geography. *The American Journal of Sociology, 14*(3), 371–407.

Hiday, V. A. (2009, November). A tribute to Amos Hawley: Mentor, colleague and friend. *Footnotes of the American Sociological Association, 37*(8). http://www.asanet.org/footnotes/nov09/tribute_1109.html

Hipp, J. R. (2007). Block, tract, and levels of aggregation: Neighborhood structure and crime and disorder case and point. *American Sociological Review, 72*(5), 659–680.

Hirshman, C. (2009, November). A tribute to Amos Hawley: Mentor, colleague and friend. *Footnotes of the American Sociological Association, 37*(8). http://www.asanet.org/footnotes/nov09/tribute_1109.html

Howell, F. M. (2004). *Spatial analysis in rural sociology*. Presented at the Annual Meetings of the Rural Sociological Society, Sacramento, CA.

Howell, F. M., & Porter, J. R. (2010). Surveys and GIS in The Handbook of Survey Research, 2nd Edition. P. Marsden and J. Wright (Eds.). United Kingdom: Emerald Press.

Jefferson, M. (1931). Distribution of the world's city folks. *Geographical Review, 21*, 446–465.

Johnson, J. H. (1967). *Urban geography: An introductory analysis*. Oxford: Pergamon Press.

King, G. (2002). Isolating spatial autocorrelation, aggregation bias, and distributional violations in ecological inference: Comment on Anselin and Cho. *Political Analysis, 10*, 298–300.

Knorr-Cetina, K. (1983). The micro-sociological challenge of macro-sociology: Towards a reconstruction of social theory and methodology. In K. Knorr-Cenina & A. V. Cicourel (Eds.), *Advances in social theory and methodology*. Boston: Routledge & Kegan Paul.

Kuhn, T. (1962). *The structure of scientific revolutions*. London: University of Chicago Press.

Land, K. C. (1990). Structural covariates of homicide rates: Are there any invariances across time and social space? *The American Journal of Sociology, 95*(4), 922–963.

Lawson, A. B. (2009). *Bayesian disease mapping: Hierarchical modeling in spatial epidemiology*. Boca Raton: Chapman and Hall.

Levine, R. F. (Ed.). (2004). *Enriching the sociological imagination: How radical sociology changed the discipline*. Boston: Brill.

Lichter, D. T., & Fuguitt, G. V. (1982). The transition to nonmetropolitan population deconcentration. *Demography, 19*, 211–221.

Lichter, D. T., & Fuguitt, G. V. (1989). Small town growth and population dispersal. In G. V. Fuguitt, D. L. Brown, & C. L. Beale (Eds.), *Rural and small town America* (pp. 63–104). New York: Russell Sage.

Lieberson, S., & Lynn, F. B. (2002). Barking up the wrong branch: Scientific alternatives to the current model of sociological science. *Annual Review of Sociology, 28*, 1–19.

Lobao, L., & Saenz, R. (2002). Spatial Inequality and Diversity as an Emerging Research Area. *Rural Sociology, 67*, 497–511.

Lobao, L. M. (1990). *Locality and inequality: Farm and industry structure and socioeconomic conditions*. Albany: State University of New York Press.

Lobao, L. M., Hooks, G., & Tickameyer, A. (2007). *The sociology of spatial inequality*. Albany: State University of New York Press.

Lobao, L. M., Rulli, J., & Brown, L. A. (1999). Macrolevel theory and local-level inequality: Industrial structure, institutional arrangements, and the political economy of redistribution, 1970–1990. *Annals of the Association of American Geographers, 89*(4), 571–601.

Losch, A. (1937). Population cycles as a cause of business cycles. *Quarterly Journal of Economics, 51*(4), 649–662.

Losch, A. (1939). In Stolper W. F (1954, Ed.), *The economics of location*. New Haven: Yale University Press.

Marshall, J. U. (1969). *The location of service towns: An approach to the analysis of central place systems*. Toronto: University of Toronto Press.

McDonald, L. (1993). *The early origins of the social sciences*. Montreal: McGill-Queen's University Press.

Melvin, B. L. (1927). Methods of social research. *The American Journal of Sociology, 33*(2), 194–210.

Merton, R. K. (1968). *Social theory and social structure*. New York: Free Press.

Messner, S. F., & Anselin, L. (2004). Spatial analyses of homicide with areal data: Chapter 6. In M. F. Goodchild & D. G. Janelle (Eds.), *Spatially integrated social sciences*. Oxford: Oxford University Press.

Messner, S. F., Anselin, L., Baller, R., Hawkins, D. F., Deane, G., & Tolnay, S. E. (1999). The spatial patterning of county homicide rates: An application of exploratory spatial data analysis. *Journal of Quantitative Criminology, 15*(4), 423–450.

Michelson, W. (1976). *Man and His Urban Environment: A Sociological Approach*. Reading, Mass: Addison-Wesley. 273 pp. rev. ed.

Mills, C. W. (1959). *The sociological imagination*. Oxford: Oxford University Press.

Moore, B. (1920). The scope of ecology. *Ecology, 1*, 3–5.

Moran, P. A. P. (1950). Notes on Continuous Stochastic Phenomena. *Biometrika 37*(1), 17–23.

Morenoff, J. D., Sampson, R. J., & Raudenbush, S. W. (2001). Neighborhood inequality, collective efficacy, and the spatial dynamics of urban violence. *Criminology, 39*(3), 517–558.

Mouw, T., & Entwisle, B. (2006). Residential segregation and interracial friendship in schools. *The American Journal of Sociology, 112*(2), 394–441.

Mowrer, E. R. (1938). The isometric map as a technique of social research. *The American Journal of Sociology, 44*(1), 86–96.

National research Council (NRC). (1997). *Rediscovering geography: New relevance for science and society*. Washington, DC: National Academy Press.

Ormsby, T., Napoleon, E., Burke, R., Groess, C., & Feaster, L. (2001). Getting to Know ArcGIS Desktop. *Basics of ArcView, ArcInfo, and ArcEditor*. Redlands, California: ESRI.

O'Sullivan, D., & Unwin, D. (2003). *Geographic information analysis*. Hoboken: Wiley.

O'Sullivan, D., & Unwin, D. (2004). *Geographic Information Analysis*. Hoboken, NJ: John Wiley & Sons, Inc.

Odum, E. P., & Barrett, G. W. (2005). *Fundamentals of ecology*. Florence: Brooks/Cole.

Okabe, A., Satoh, T., Furuto, T., Suzuki, A., & Okano, K. (2008). Generalized network Voronoi diagrams: Concepts, computational methods, and applications. *International Journal of Geographical Science, 22*, 965–994.

Olson, R. (1993). *The emergence of the social sciences 1642–1792*. New York: Twayne Publishers.

Osgood, D. W. (2000). Poisson-based regression analysis of aggregate crime rates. *Journal of Quantitative Criminology, 16*(1), 21–43.

Park, R. E. (1926). The urban community as a spatial pattern and a moral order. In E. W. Burgess (Ed.), *The urban community*. Chicago: University of Chicago Press.

Park, R. E. (1929a). Urbanization as measured by newspaper circulation. *The American Journal of Sociology, 35*(1), 60–79.

Park, R. E. (1929b). Sociology, community and society. In *Research in the social sciences*. New York: Macmillan Co.

Park, R. E. (1934). Dominance. In M. Roderick (Ed.), *Readings in human ecology*. Ann Arbor: University of Michigan Press.

Park, R. E. (1936a). Human ecology. *The American Journal of Sociology, 42*, 3–49.

Park, R. E. (1936b). Succession, an ecological concept. *American Sociological Review, 1*, 171–179.

Park, R. E. (1939). Symbiosis and socialization: A frame of reference for the study of society. *The American Journal of Sociology, 45*, 1–25.

Park, R. E. (1952). In E. Hughes, C. Johnson, J. Masuoka, R. Redfield, & L. Wirth (Eds.), *Human communities: The city and human ecology*. Glencoe: The Free Press.

Park, R. E., & Burgess, E. W. (1921). *Introduction to the science of society*. Chicago: University of Chicago Press.

Park, R. E., & Burgess, E. W. (1925). This is the title of the book and the editors simple pulled together the work for the book, thus no chapter name. *The city*. Chicago: University of Chicago Press.

Paulsen, D. J., & Robinson, M. B. (2004). *Spatial aspects of crime: Theory and practice*. Boston: Pearson Education, Inc.

Peters, A., & MacDonald, H. (2005). *Unlocking the census with GIS*. Redlands, CA: ESRI Press.

Polston, D. L. (2009, November). A tribute to Amos Hawley: Mentor, colleague and friend. *Footnotes of the American Sociological Association, 37*(7). http://www.asanet.org/footnotes/nov09/tribute_1109.html

Porter, J. R. (2004). *The relationship between political cynicism and right-wing authoritarianism: A case study of the Louisville-Jefferson county government merger*. M.A. thesis, University of Louisville, Louisville.

Porter, J. R. (2008). *The spatial demography of reported crime: An examination of urban-rural crime articulation and associated spatio-temporal diffusion processes, Washington, DC. 1990–2000*. Ph.D. dissertation, Mississippi State University, Starkville. UMI/Proquest Publishers.

Porter, J. R. (2010). *Tracking the mobility of crime: New methodologies and geographies in modeling the diffusion of crime*. Newcastle: Cambridge Scholars Publishing.

Porter, J. R. (2011). Context, location, and space: The continued development of our geo-sociological imaginations. *The American Sociologist, 42*, 288–302.

Porter, J. R., & Howell, F. M. (2009). On the 'Urbanness' of metropolitan areas: Testing the homogeneity assumption, 1970–2000. *Population Research and Policy Review, 28*, 589–613.

Ralston, B. A. (2004). *GIS and public data*. Clifton Park NY, Delmar Learning.

Raudenbush, S. W., & Robert, J. S. (1999). Ecometrics: Toward A Science of Assessing Ecological Settings, with Application to the Systematic Social Observation of Neighborhoods. *Sociological Methodology 29*, 1–41.

Raudenbush, S. W., & Bryk, A. S. (2002). *Hierarchical linear models: Applications and data analysis methods* (2nd ed.). Thousand Oaks: Sage.

Robinson, W. S. (1950). Ecological correlations and the behavior of individuals. *American Sociological Review, 15*, 351–357.

Rosenberg, C. E. (1962). *The cholera year*. Chicago: The University of Chicago Press.

Sampson, R. J. (2002). Transcending tradition: New directions in community research, Chicago style. *Criminology, 40*(2), 213–230.

Sampson, R. J. (2003). The neighborhood context of well-being. *Perspectives in Biology and Medicine, 46*(3), S53–S64.

Sampson, R. J. (2008). Moving to inequality: Neighborhood effects and experiments meet social structure. *The American Journal of Sociology, 114*(1), 189–231.

Sampson, R. J. (2009). Disparity and diversity in the contemporary city: Social (dis)order revisited. *The British Journal of Sociology, 60*(1), 1–31.

Sampson, R. J., & Raudenbush, S. W. (1999). Ecometrics: Toward a science of assessing ecological settings, with application to the systematic observation of neighborhoods. *Sociological Methodology, 29*, 1–41.

Sampson, R. J., & Sharkey, P. (2008). Neighborhood selection and the social reproduction of concentrated racial inequality. *Demography, 45*(1), 1–29.

Sampson, R. J., Raudenbush, S. W., & Earls, F. (1997). Neighborhoods and violent crime: A multilevel study of collective efficacy. *Science, 277*, 918–924.

Sampson, R. J., Morenoff, J. D., & Earls, F. (1999). Beyond social capital: Spatial dynamics of collective efficacy for children. *American Sociological Review, 64*, 633–660.
Sampson, R. J., Morenoff, J. D., & Gannon-Rowley, T. (2002). Assessing "Neighborhood Effects": Social processes and new directions in research. *Annual Review of Sociology, 28*, 443–478.
Sampson, R. J., McAdam, D., MacIndoe, H., & Weffer-Elizondo, S. (2005). Civil society reconsidered: The durable nature of community structure of collective action. *The American Journal of Sociology, 111*(3), 673–714.
Schnore, L. F. (1958). Social morphology and human ecology. *The American Journal of Sociology, 63*(6), 620–634.
Schootman, M., Jeffe, D., Kinman, E., Higgs, G., & Jackson-Thompson, J. (2004). Evaluating the utility and accuracy of a reverse telephone directory to identify the location of survey respondents. *Annals of Epidemiology, 15*(2), 160–166.
Schuyler, E., Gipson, R., & Walsh, J. (2005). *Mapping hacks: Tips & tools for electronic cartography*. Cambridge: O'Reilly.
Scott, M. L., & Mears, D. P. (2005). *Understanding and reducing survey non-response rates among rural farming populations through spatial analysis*. Presented at the National Institute of Justice Meetings, September 8, 2005, Savannah.
Shanahan, M. J., & Macmillan, R. (2008). *Biography and the sociological imagination: Contexts and contingencies*. New York: W.W. Norton and Company.
Small, A. W., & Vincent, G. E. (1894). *An introduction to the study of society*. New York: American Book Co.
Snow, J. (1949). *On the mode of communication of cholera* (2nd ed., 1855, Vols. 8) Los Angeles, CA: UCLA Press.
Sorokin, P. A. (1964). *Sociocultural causality, space, and time: A study of referential principles of sociology and social science*. New York: Russell & Russell Inc.
Sorokin, P. A., Zimmerman, C. C., & Galpin, C. J. (Eds.). (1930). *A systematic source book in rural sociology* (Vol. I). Minneapolis: University of Minnesota Press.
Steinberg, S. J., & Steinberg, S. L. (2005). *Geographic information systems GIS for applied social research*. London: Sage.
Terpenning, W. A. (1928). Requisites to rural social organization. *The American Journal of Sociology, 33*(5), 737–753.
Thomas, F. (1925). *The environmental basis of society*. New York: The Century Co.
Thomas, J. K. (1992). Agriculture, the environment and rural sociology in the south. *Southern Rural Sociology, 9*(1), 1–22.
Tickamyer, A. R. (2000). Spatial Inequality in the Future of Sociology. *Contemporary Sociology, 28*(6):805–813.
Ullman, E. L. (1941). A theory of location for cities. *The American Journal of Sociology, 46*, 853–864.
Von Thunen, J. H. (1826). In P. Hall (1966, Ed.), No Chapter Title, translated book. *Isolated state*. Oxford: Pergamon Press.
Voss, P. R. (2007a). Introduction to the special issue on spatial demography. *Population Research and Policy Review, 26*(4), 455–456.
Voss, P. R. (2007b). Demography and spatial social science. *Population Research and Policy Review, 26*(4), 457–476.
Waller, L. A., & Gotway, C. A. (2004). *Applied spatial statistics for public health data*. Hoboken: Wiley.
Weber, A. [translated by Carl J. Friedrich from Weber's 1909 book] (1929). *Theory of the Location of Industries*. Chicago: The University of Chicago Press.
Wilkinson, K. P. (1991). *The community in rural America*. New York, NY: Greenwood Press.
Wu, J., Funk, T. H., Lurman, F. W., & Winer, A. M. (2005). Improving spatial accuracy of roadway networks and geocoded addresses. *Transactions in GIS, 9*(4), 585–601.

Index

J.R. Porter and F.M. Howell, *Geographical Sociology: Theoretical Foundations and Methodological Applications in the Sociology of Location*, GeoJournal Library 105,
DOI 10.1007/978-94-007-3849-2,

G

H

MIX
Papier aus verantwortungsvollen Quellen
Paper from responsible sources
FSC® C105338

If you have any concerns about our products,
you can contact us on
ProductSafety@springernature.com

In case Publisher is established outside the EU,
the EU authorized representative is:
Springer Nature Customer Service Center GmbH
Europaplatz 3, 69115 Heidelberg, Germany

Printed by Libri Plureos GmbH
in Hamburg, Germany